# The
# GREAT NORTHERN RAILWAY
## at

# FARRINGDON

## Allan Sibley

*Above:* At Farringdon on 15th March 1949, BR(ER) class J50 (ex-GNR class J13) 0-6-0ST still carrying its LNER number 8760, at the top of the 1 in 40 climb from under the Ray Street Gridiron with a Ferme Park to Herne Hill goods train. The word 'TISH' is chalked on the smokebox door. On the left the tracks to and from Farringdon Street depot connect to the Widened Lines. *(John Aylard).*

*Front cover illustration*: In around 1900 at Metropolitan Passenger Junction at the south end of Farringdon Street station, where the connection to the London Chatham & Dover Railway via Ludgate Hill diverges from the Metropolitan Railway's Widened Lines, District Railway Beyer-Peacock 4-4-0T No. 26 approaches Farringdon on the eastbound Met/District/Circle line track from Aldersgate Street (now Barbican) with a clockwise Inner Circle passenger train, while GNR 126 series (later class F4) 0-4-2 No. 119 emerges from Snow Hill tunnel with a Down goods train. *(Painting by Jack Hill).*

*Every effort has been made to acknowledge the sources of illustrations used in this publication. Any errors or oversights which may have occurred are inadvertent and will be corrected in subsequent issues providing notification is sent to the Editor at editor@gnrsociety.com*

Published by the Great Northern Railway Society     gnrsociety.com
Printed by David J Richards Ltd, 1 West Park Street, Chatteris PE16 6AH     djrprint@chessmail.co.uk
**1st Edition 2019     ISBN 978-1-912113-65-1**

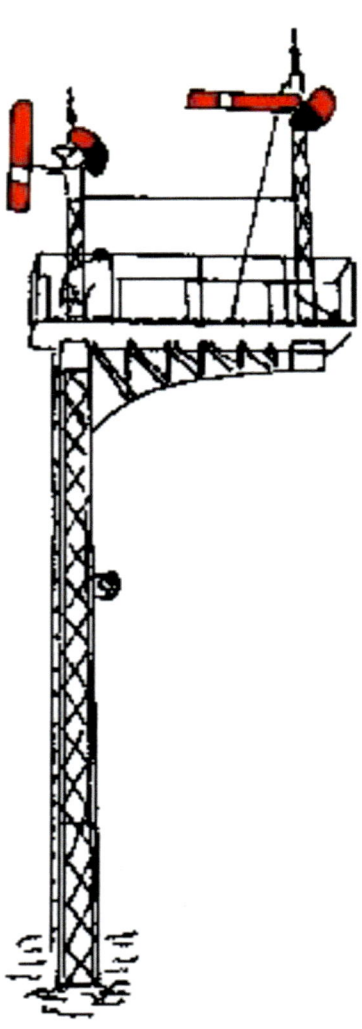

# Contents

# 1 Preface

This origins of this book go back to mid-2013 when Jonathan David, the then Editor of the Historical Model Railway Society's newsletter POINTS, asked if I could provide some captions to a selection of photographs of Farringdon goods depot that he had submitted for publication to the then Editor of the HMRS JOURNAL, Adam Muir.

I thought I was reasonably familiar with the subject and so began writing but it soon became clear that there was an interesting story waiting to get out of the captions. The resultant article ran to 22½ quarto-size pages which appeared in the January-March 2014 HMRS JOURNAL and I was modestly pleased with the favourable comments received from HMRS and GNRS members. However, some photographs had had to be omitted from the JOURNAL for space reasons. Also, as a result of help from a few HMRS members, further information and photos came to light.

Thus an enlarged article appeared in serialised form during 2016 in GREAT NORTHERN NEWS issues 205 to 209. This book is a re-combination of those five parts plus additional material regarding the nearby Turnmill Street Stables and the railway's use of horse power.

Although this book appears under my name I am grateful for the assistance of several fellow members of the GNRS, as noted in the 'Acknowledgements'. However any errors of fact and typographical mistakes are entirely down to me.

While the story of the Metropolitan Railway's conception and construction has already been told in numerous books and magazine articles, it is briefly described here for completeness and especially for the benefit of readers who are unfamiliar with its history. However, I have not included the post-1977 developments (Thameslink and Crossrail) which are left to future historians to record.

Allan Sibley, March, Cambridgeshire.
February 2019.

# 2 Introduction

One of the fascinating features of London's railways up to the start of the Second World War was the number of small goods and coal depots packed into an intensely urban area. For example, in the first half mile out from Fenchurch Street there were six: Goodmans Yard (GER), Royal Mint Street (GNR), Haydon Square (L&NWR), City Goods (Midland), East Smithfield (GER) and Commercial Road (LT&SR).

The German bombing of London's East End during the Blitz destroyed some depots and badly damaged others but several survived into the 1960s and a few of their abandoned structures lasted into the 21st century. Patterns of traffic changed dramatically after the war as manufacturing industry left the central London area. The dye works, foundries, snuff factories, tanneries and slaughterhouses closed one by one and with them went the need for raw materials and fuel. Simultaneously the Clean Air Act of 1956 (a direct result of London's "Great Smog" of 1952) and subsequent legislation led to a drastic reduction in the hitherto huge appetite of the metropolis for domestic coal. Add to that the increasing use of long-distance road transport to markets such as Smithfield (meat), Billingsgate (fish) and Covent Garden (flowers, fruit and vegetables) and it is easy to see the causes of the demise of many inner-city railway goods yards.

I can still remember as a eight to twelve year old trainspotter, looking eagerly out of the carriage window hoping to spot a shunting loco or two as I passed Bishopsgate on the approach to Liverpool Street. Farringdon was another of my "could be a lurking shunter" locations. Clearly visible from passing Metropolitan and Circle line trains, the adjacent goods depot seemed to tower menacingly over the passenger station. Squeezed between Farringdon Road and the ex-GNR "Widened Lines" it looked incredibly thin for its height (at the north end). Views of it were necessarily brief and limited so I did not know at the time that it had already been closed for a couple of years.

It was therefore not surprising that I never saw a Hornsey N1, J50 or J52 busying itself there. Likewise I did not realise then that Farringdon was an "iceberg" of a depot: what was visible above ground from passing trains was but a fraction of the total extent of the facilities.

This is possibly the best point to try to clarify the nomenclature.

According to Hilliam*, Farringdon Street took its name from Farringdon Ward which in turn was named after two 13th century aldermen of the City of London, William and Nicholas de Farndon. In pre-railway days the thoroughfare ran from Blackfriars Bridge to the junction with Holborn Hill and Skinner Street. Up to about 1860 the continuation north was called Victoria Street but after this road was widened and straightened, as well as having the railway laid under it, it became Farringdon Road. At first the Metropolitan Railway passenger station was named after Farringdon Street but subsequently became Farringdon & High Holborn, then plain Farringdon. The GNR's goods depot has officially been known as both Farringdon Street and Farringdon Road from its planning in the 1890s to closure in the 1950s, but it's other title, sometimes also used 'officially', was City Goods Station, despite it not being within the City of London. However, the Midland Railway's "City Goods" about half a mile away at Whitecross Street, was within the City boundary as were both the GN's and Midland's other 'city' goods depots, side by side at Royal Mint Street.

The other unresolved anomaly is the road running past the entrance to the passenger station. It seems to have been an eastward portion of Charles Street until the arrival of the railway but then became Cow Cross Street. From that time on it was shown on various maps as either Cow Cross or Cowcross Street, a name also used (e.g., figs. 1, 4 and 10) for part of Turnmill Street.

\* see chapter 15 'References and Acknowledgements'
Hopefully that clarifies the situation.

# 3  Farringdon Street and Farringdon Road Before The Railway

On completion of the first Blackfriars Bridge over the River Thames in 1769 the then new Corporation of the City of London set about improving the road connections to its northern end. The first section of new road, from there to Ludgate Circus (Fleet Street / Ludgate Hill) was named New Bridge Street and from Ludgate Circus northward it was Farringdon Street which up to the 1850s ended at a staggered junction with Holborn Hill and Skinner Street which were replaced in 1869 by the elevated Holborn Viaduct.

The map (fig.1) right is dated 1862 but shows the "Metropolitan Railway Terminus" which did not open until 10th January of the following year. It is on the western side of the cutting, nestled up against the Farringdon Road retaining wall. However the map also shows it extending south of Charles Street. Two tracks emerge from the south end and follow almost the exact course of the LC&DR line from Ludgate Hill to Farringdon Street which was not built until 1866, after this station had closed. Its replacement was built alongside and in fact the LC&DR when built ran on a slightly more easterly alignment than shown on this map. The land on the western side of Farringdon Road is shown as unoccupied by buildings but it did not remain thus for very long.

It is interesting to note that the name Cow Cross Street appears in the middle of Turnmill Street (to the east [right] of the station) whereas it was later the name of the section of Charles Street to the east of Farringdon Street/ Road. Another map (fig. 2 overleaf) is dated 26th February 1851, updated to 1st January 1860. It was probably prepared in conjunction with work on the Great Northern Railway's land at King's Cross which the cartographer has coloured red. It shows the roads immediately north of the Holborn Hill / Skinner Street junction as Victoria Street, which ran into Coppice Row and then Guildford Place. In the period from about 1852 until 1861 these roads were cleared

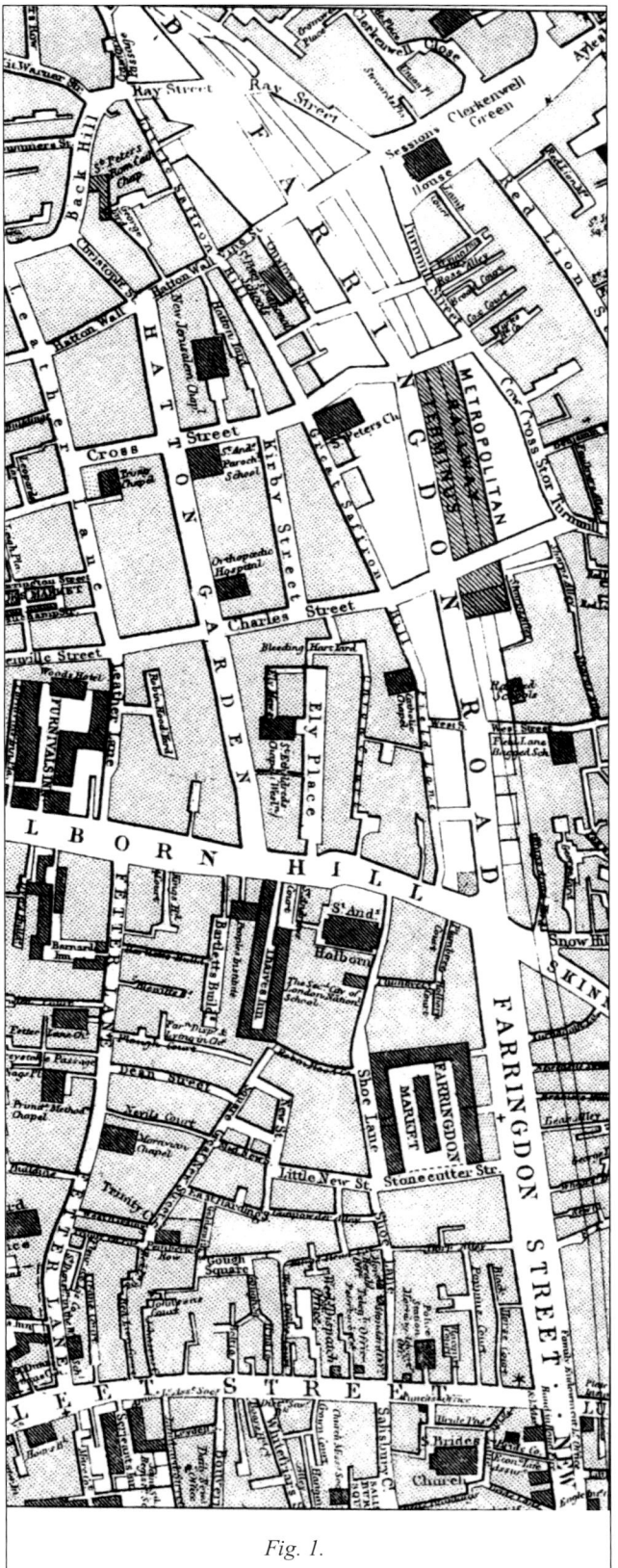

and Farringdon Road built on their alignment northward from Holborn Hill / Skinner St.

Work was delayed by financial problems, the need for clearance of existing properties and uncertainty regarding the route of the then proposed Metropolitan Railway. It had progressed as far north as Ray Street by 1862 and completion from there to Lloyd Baker Street was undertaken at the same time as the railway construction so that the latter could run under the road. North of Lloyd Baker Street, the route continued as Bagnigge Wells Road and Hamilton Road. In 1863 these all became King's Cross Road, joining Pentonville Road a few hundred yards to the east of King's Cross. The contractor was John Jay, well known for his work at King's Cross and elsewhere.

The Metropolitan Railway was the world's first underground railway, running from Bishop's Road, Paddington to an eastern terminus at Farringdon Street station located on the eastern side of Farringdon Road. Opened on 10th January 1863 and three and three-quarter miles long, it was built by the 'cut and cover' method almost entirely under the "New Road". This had been constructed from the 1750s onward and comprised (from west to east) Praed Street, Marylebone Road and Euston Road. Having an end-on connection with the then broad gauge Great Western Railway at Paddington, the Metropolitan was laid with mixed gauge track and worked for a short time using GWR broad gauge locomotives and rolling stock. It was steam-worked from the outset until electric traction replaced steam on passenger trains in 1905.

A Royal Commission of 1846 had prohibited railways from entering the City of London and West End, hence the line did not encroach into the area south of the New Road and terminated at the western boundary of the City. However it was not long before this restriction was lifted and the extension built into the City of London at Moorgate Street.

*Right: Fig. 2. Map dated 26th February 1851, updated to 1st January 1860. King's Cross station is at the top left and the area of land occupied by the goods depot is remarkably large. Note too the absence of St Pancras station, still some eight years away. The road that runs in front of the Great Northern Hotel is Weston Place while Pancras Place (not Pancras Road at that time) is but a short thoroughfare between Weston Place and Church Row.*

*Blackfriars Bridge is at bottom right. Farringdon Street exists but to its north Farringdon Road has yet to be built over what was then Victoria Street, Coppice Row and Guildford Place.*

*In 1863 Bagnigge Wells Road and Hamilton Road became King's Cross Road, between Farringdon Road and Pentonville Road..*

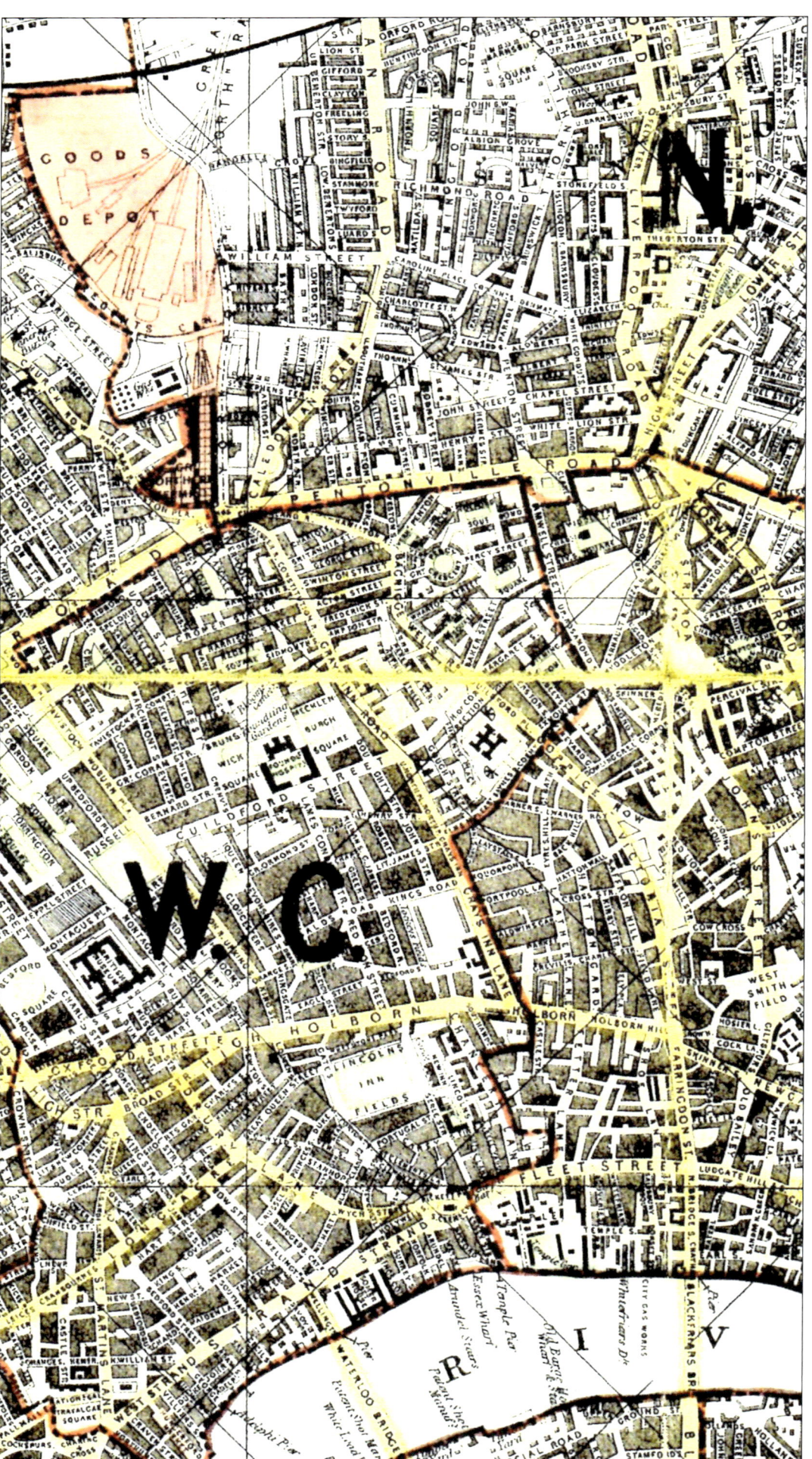

*Fig. 3. Extract from (a rather feint copy of) the 1871 Ordnance Survey map showing Ray Street and the north end of Farringdon Road at bottom right, with the railway from King's Cross coming in from the top left. The route of the second Clerkenwell tunnel is shown, marked "Metropolitan Railway" to the north-east of Farringdon Road which has the original Metropolitan tracks beneath it. Nothing to do with the railway but note the huge area of the "Middlesex House of Correction", also known as Coldbath Fields Prison, the details of which were not fully shown for security reasons. It closed in 1885 and from 30th August 1889 the buildings became the home of the GPO's Mount Pleasant Sorting Office (still operational but much altered). Nearby and to the east is the "Middlesex House of Detention". (Alan Godfrey edition London Sheet 50).*

*Fig. 4. Extract from (a rather feint copy of) the 1873 Ordnance Survey map showing the second Farringdon station (top centre) on the right hand side of the cutting, after the demolition of the first station but before construction of the goods depot. There are no buildings close to Charterhouse Street, Holborn Viaduct or the west side of Farringdon Road. The station at the bottom centre is the LC&DR's Ludgate Hill. (Alan Godfrey edition London Sheet 62)*

The Great Northern Railway opened its permanent King's Cross passenger terminus in 1852, the original temporary Maiden Lane station becoming part of the goods depot. However the company had ambitions to serve the City of London. After the opening of the Metropolitan Railway the GN directors realised that their best option was to make a physical connection with the Met's route at King's Cross. Thus a line was built in tunnel from the King's Cross 'throat' to join the eastbound Metropolitan and another from the westbound track via a sharp curve on a rising gradient in tunnel under the Great Northern Hotel, emerging on the west side of King's Cross station. Through services between the GN and Farringdon Street commenced on 1st October 1863, less than nine months after the Metropolitan services had begun.

The Metropolitan subsequently extended to Aldersgate Street and Moorgate Street from a new Farringdon Street station built alongside the first and opened on 23rd December 1865.

On 1st January 1866 the London, Chatham & Dover Railway opened a short extension from its new Ludgate Hill station north of the Thames to an end-on junction with the Metropolitan extension from Farringdon Street at West Street Junction. The first Farringdon Street station closed on 1st March 1866. The GNR started a through passenger service to Ludgate Hill and Victoria. Further destinations south of the Thames were added in future years. Meanwhile LC&DR trains worked through as far north as Hatfield. Finally (for the moment) the Metropolitan opened a further extension from Moorgate Street to Liverpool Street on 12th July 1875 although confusingly for passengers using the GER terminus, the Metropolitan station was named Bishopsgate until 1909.

The two track arrangement between King's Cross and Moorgate Street soon proved inadequate as traffic grew. The Metropolitan Railway (Additional Powers) Act of 25th July 1864 authorised the company to double the capacity and initially additional lines were provided between Farringdon Street and Moorgate Street.

The quadrupling between King's Cross and Farringdon Street was more complicated. It requiring a duplicate Clerkenwell Tunnel (bored, not "cut and cover" as the first one was), new junctions at King's Cross and a "dive-under" just north of Farringdon Street to take the new tracks from the north side of the Met to the south side via what became known as the "Ray Street Gridiron". The gradients either side were quite severe: that on the Farringdon side being about 270 yards at 1 in 40. To enable the construction of the "Widened Lines" (sometimes referred to as the "City Widened Lines") to be carried out, the GN suburban service was suspended from 1st July 1867 to 1st March 1868.

The work included yet another line joining the Metropolitan in tunnels just west of King's Cross, this being the Midland Railway from St Paul's Road Passenger Junction, near Kentish Town. Thus on 13th July 1868 Moorgate Street became the first London terminus to be used by the Midland Railway (St Pancras did not open until the following 1st October). Before too long the Midland too was running through trains from Bedford to Moorgate Street and Finchley Road to Victoria.

Although the Metropolitan was primarily a passenger railway, proposals for a goods depot at Farringdon were first put forward as early as 1864 by one William Burchell Jnr, son of William Snr, the Metropolitan's solicitor. The scheme was floated the following year by the Metropolitan

Railway Warehousing Company Ltd but it was unsuccessful and the company was wound up in 1872.

The Great Western Railway had running powers over the Metropolitan and on 3rd May 1869 opened the first goods depot in the area - between Farringdon Street and Aldersgate Street stations, below the then new Smithfield Meat Market.

Having succeeded in introducing its passenger service from its suburbs to the City of London, the Great Northern Railway began to consider its London goods traffic. There was a lucrative trade to be had in the City but King's Cross goods depot was somewhat remote and road cartage relatively expensive mainly due to the number of horses involved. The Company already had a City depot from 1858 at Royal Mint Street but it was small, cramped and almost as remote as King's Cross, being about ¼ mile east of Fenchurch Street. Access was by a circuitous route from Finsbury Park via Canonbury, Bow and Stepney. Thus the plot of land at Farringdon, vacant since the removal of the Metropolitan's first (terminus) station, looked to be an attractive site for the 'City Goods Station' as it was sometimes referred to. The GN leased this land from the Metropolitan in 1873.

A full chronology of the development of the railways here is given on page 10 and diagrams are on pages 11/12.

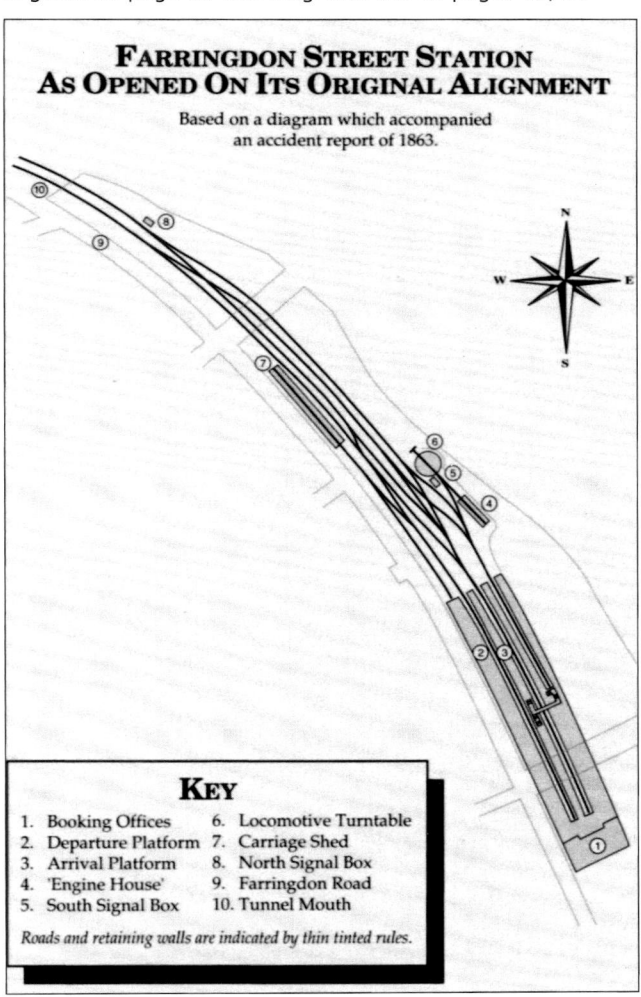

**FARRINGDON STREET STATION AS OPENED ON ITS ORIGINAL ALIGNMENT**

Based on a diagram which accompanied an accident report of 1863.

**KEY**

1. Booking Offices
2. Departure Platform
3. Arrival Platform
4. 'Engine House'
5. South Signal Box
6. Locomotive Turntable
7. Carriage Shed
8. North Signal Box
9. Farringdon Road
10. Tunnel Mouth

*Roads and retaining walls are indicated by thin tinted rules.*

*Fig. 5. From* STEAM OF THE WIDENED LINES *volume 2 by Geoff Goslin (Connor & Butler 1998) reproduced with the permission of Jim Connor. Note the turntable (6) which only lasted about three years, yet the recess in the retaining wall and the massive iron girder above it remain* in situ. *(Jim Connor).*

# 5  A Chronology of the Widened Lines 1852 - 1979

| | |
|---|---|
| 14.10.1852 | King's Cross (Great Northern Railway) main line terminus station opened. |
| 10.01.1863 | Paddington to Farringdon Street terminus station (Metropolitan Railway) opened. |
| 01.10.1863 | King's Cross (GNR) to King's Cross (Met) and King's Cross (Met) first station opened. |
| 21.12.1864 | { Ludgate Hill station (LC&DR) opened on a temporary site.<br>{ Blackfriars Bridge to Ludgate Hill (London Chatham & Dover Railway) opened. |
| 01.06.1865 | Ludgate Hill new station (LC&DR) opened on permanent site. |
| 23.12.1865 | Farringdon Street new through station on adjacent site to terminus, and extension to Moorgate Street station (Met) via Aldersgate Street station opened. |
| 01.01.1866 | { Ludgate Hill (LC&DR) station to West Street Junction (GNR) opened.<br>{ Aldersgate Street to Snow Hill Junction (The Smithfield Curve) opened.<br>{ King's Cross York Road station (GN) opened. |
| 01.03.1866 | Farringdon Street (Met) terminus station closed. |
| 13.07.1868 | St Paul's Road Passenger Junction (Midland Railway) to King's Cross (Met) opened. |
| 01.10.1868 | St Pancras main line station (Mid) opened. |
| 03.05.1869 | Smithfield goods station (GWR) opened. |
| 02.03.1874 | Holborn Viaduct station (LC&DR) opened. |
| 01.08.1874 | Snow Hill station (LC&DR) opened. |
| 02.11.1874 | Farringdon Street goods station (GNR) opened. |
| --. --.1878 | King's Cross York Road (GN) station resited. |
| 01.11.1878 | Whitecross Street goods station (Mid) goods station opened. |
| 01.10.1885 | Blackfriars Bridge station closed (and subsequently incorporated into adjacent goods station). |
| 10.05.1886 | St Paul's station (LC&DR) opened. |
| 01.11.1887 | St Pancras New Goods Station (Mid) opened. |
| 01.08.1892 | St Pancras New Goods Station (Mid) renamed Somers Town goods station. |
| --.11.1894 | Farringdon Street goods station (GNR) – enlarged facilities opened. |
| 01.11.1909 | { Vine Street goods station (Met) opened.<br>{ Gower Street station (Met) renamed Euston Square. |
| 01.11.1910 | Aldersgate Street station (Met) renamed Aldersgate. |
| 01.05.1912 | Snow Hill (LC&DR) station renamed Holborn Viaduct Low Level. |
| 01.04.1916 | Aldersgate Street to Snow Hill Junction (The Smithfield curve) closed. |
| 01.06.1916 | Ludgate Hill (LD&CR) to West Street Junction (GNR) and Holborn Viaduct station (LC&DR) closed to passenger services. |
| 26.01.1922 | Farringdon Street passenger station renamed Farringdon & High Holborn. |
| --. --.1923 | Aldersgate station renamed Aldersgate & Barbican. |
| --. --.1925 | King's Cross (Met/Circle) first station renamed King's Cross & St Pancras. |
| 03.03.1929 | Ludgate Hill station (LC&DR) closed. |
| --. --.1933 | King's Cross & St Pancras (Met/Circle) first station renamed King's Cross St Pancras (ampersand deleted). |
| 01.04.1936 | Whitecross Street goods station (Mid) closed. |
| 21.04.1936 | Farringdon & High Holborn passenger station renamed Farringdon. |
| 01.07.1936 | Vine Street goods station (Met) closed. |
| 01.02.1937 | St Paul's station (LC&DR) renamed Blackfriars. |
| 14.03.1941 | King's Cross St Pancras (London Transport Passenger Board) second station opened. |
| 16.01.1956 | Farringdon Street goods station (GNR) closed. |
| 30.07.1962 | Smithfield goods station (GWR) closed. |
| 01.12.1968 | Aldersgate & Barbican station renamed Barbican. |
| 24.03.1969 | Ludgate Hill (LC&DR) to West Street Junction (GNR) closed to goods services. |
| 04.06.1973 | Somers Town goods station (Mid) closed. |
| 08.11.1976 | King's Cross York Road station (GN) temporarily closed. |
| 31.01.1977 | King's Cross York Road station (GN) reopened. |
| 05.03.1977 | { King's Cross (GN) York Road station and platform 16 both closed and their connections to/from<br>{ The Widened Lines severed. |
| 14.05.1979 | King's Cross St Pancras (Met) Widened Lines first (KX) station platforms closed. |

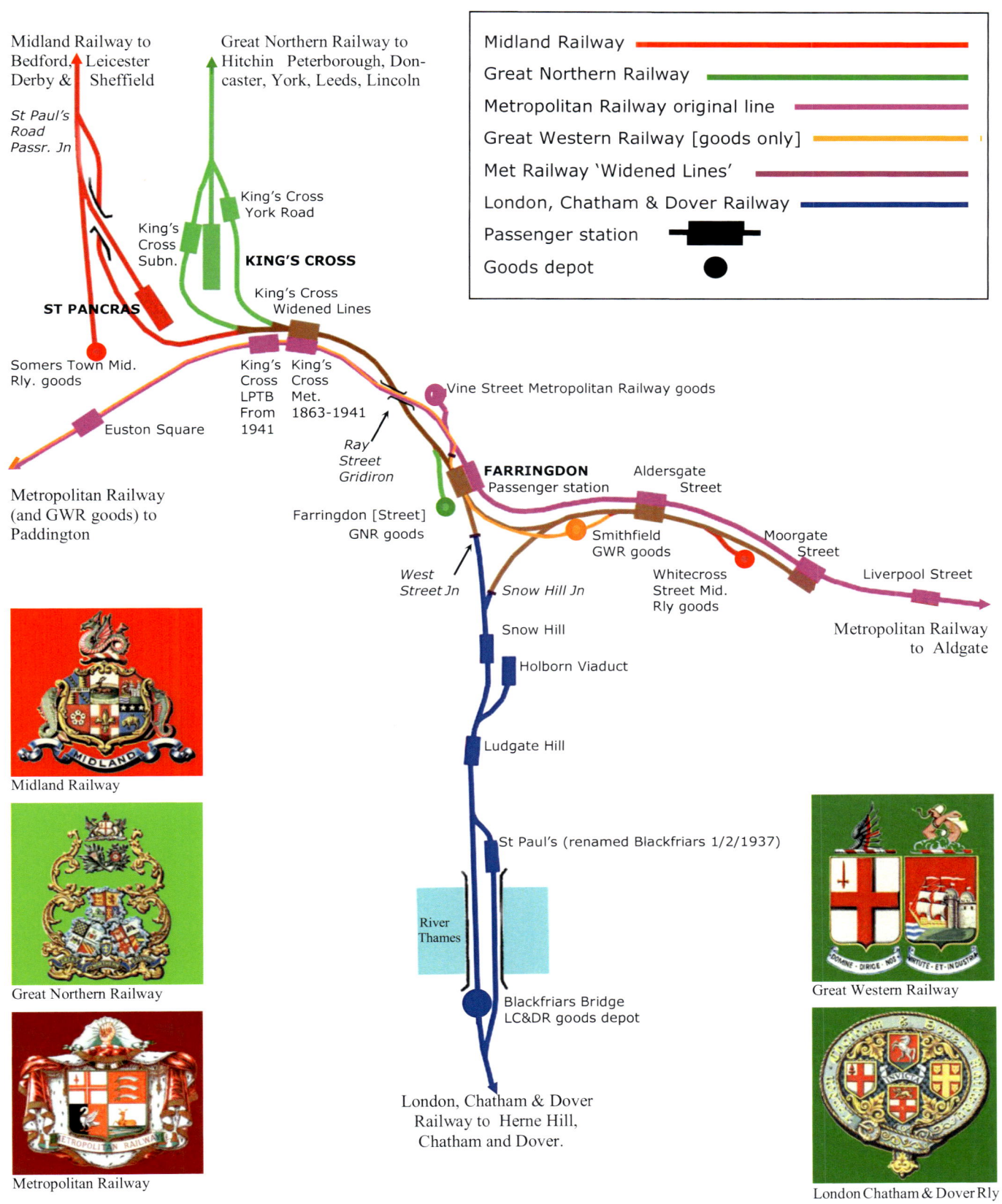

Fig. 6. Diagram of the railways around Farringdon circa 1910. (Allan Sibley)

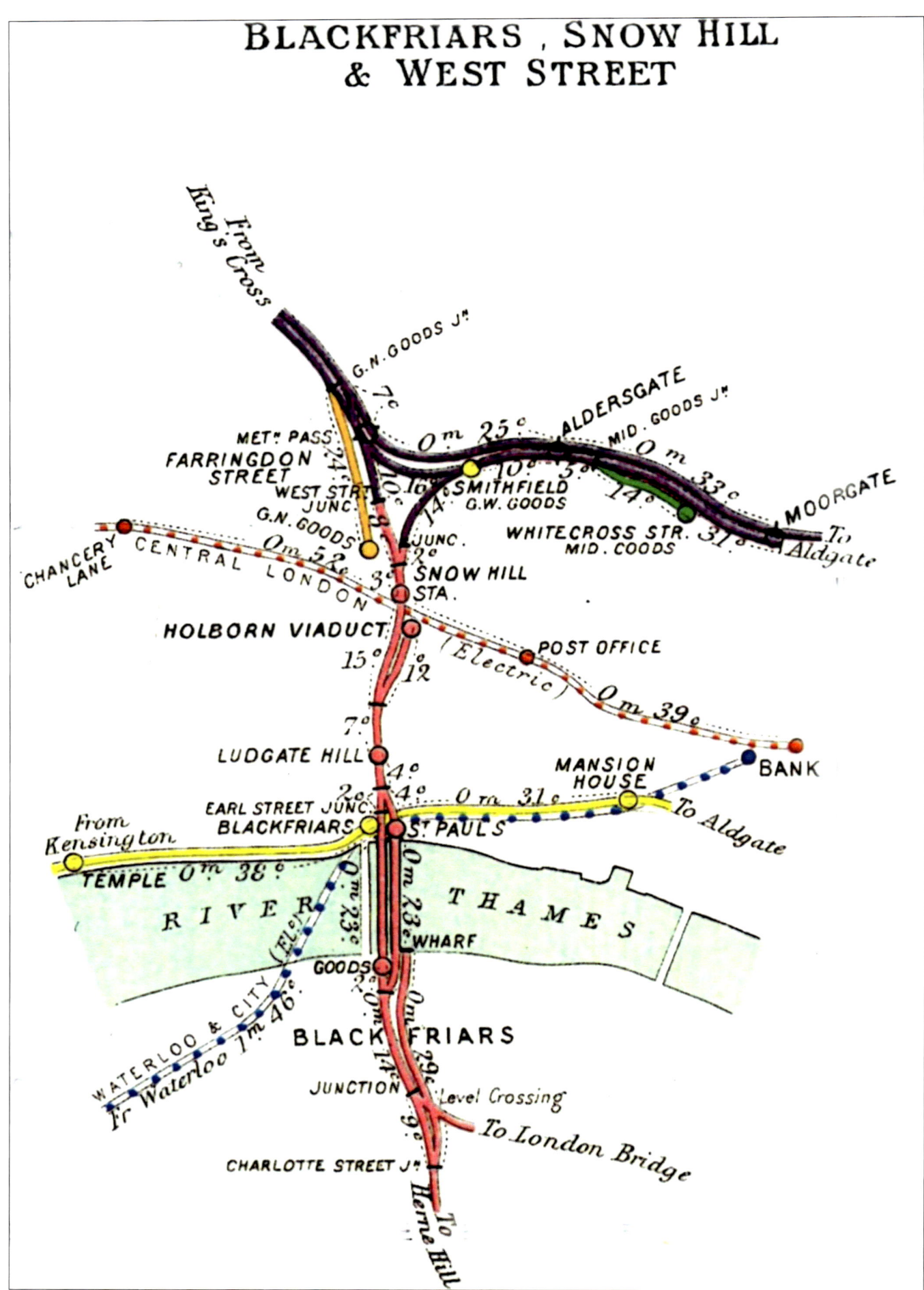

# BLACKFRIARS , SNOW HILL
# & WEST STREET

*Fig. 7. 1903 diagram from The Railway Clearing House Pre-Grouping Railway Junction Diagrams (facsimile reprint, Ian Allan, undated)*

# 6 The Passenger Stations

Before continuing with the post-1873 developments at Farringdon goods depot we must back-track slightly to look at the first (terminus) and second (through) passenger stations. As already noted the former was tucked against the west side of the cutting that had been excavated for it, as seen in fig. 5 on page 9. The engravings below, from an 1863 edition of the ILLUSTRATED LONDON NEWS, give a good idea of the arrangement although the artist has had a problem depicting the 'mixed gauge' track in fig. 8!

*Fig. 8. Looking north from under the overall roof. Farringdon Road is behind the parapet of the retaining wall on the left. Note the Turnmill Street parapet on the right with, adjacent to the chimney, a plate girder over the semi-circular 'inset' of the turntable into the retaining wall (see fig. 27, page 24). (ILLUSTRATED LONDON NEWS)*

*Fig. 9. A remarkably detailed view displaying the artist's excellent eye for perspective (remember it is an engraving, not a photograph). Turnmill Street is in the foreground and Farringdon Road on the opposite side of the cutting. The nearest bridge is on the alignment of the later Clerkenwell Road but at the time of the engraving was the culverted River Fleet or a relief channel therefor. Next is Vine Street, then the arch of Ray Street and just visible in the distance, the mouth of Clerkenwell Tunnel. Note the undeveloped land on the west side of Farringdon Road (extreme left of the picture) – it did not stay that way for long! (ILLUSTRATED LONDON NEWS)*

*Fig. 10. Another section of the 1873 OS map (see fig. 4) shows the 1865 station on the east side of the cutting and the land between it and Farringdon Road vacated by the demolition of the 1863 terminus. The station building is on the overbridge. The Widened Lines are in the left (west) platforms, the Metropolitan tracks in the right (east).*

*Fig 11. The ILLUSTRATED LONDON NEWS depicted the frontage of the new station in a contemporary issue.*

*Fig.12. A photograph from the London Transport Museum collection showing the station shortly before it was rebuilt in 1923. The two unobscured windows to the left of the canopy proclaim "BOOKING OFFICE – GREAT EASTERN, LONDON CHATHAM & DOVER AND SOUTH EASTERN RAILWAYS" and "BOOKING OFFICE – GREAT WESTERN, GREAT NORTHERN AND MIDLAND RAILWAYS" respectively. Six companies serving one station! The sign on the end of the canopy informs of:*

**FARRINGDON Sᵀ Station**

ELECTRIC TRAINS
TO ALL PARTS OF LONDON

*while that on the roof above the "Metropolitan Railway" advertises*

RESTAURANT

*although how visible it was from street level is questionable.*

*Fig.13 (above). From an engraving in the* ILLUSTRATED LONDON NEWS, *27th February 1868. The 0-4-2 is on the GNR's track but so far as I am aware the company had no locomotives that looked like this. In fact it seems that apart from the 10 months from 1st October 1863 to 1st August 1864 during which the GNR hired 0-4-2s and 0-6-0s to the Metropolitan, no tender locomotives were used on the Widened Lines. It is heading an Up train from King's Cross towards Farringdon. A 'Met' train travelling towards King's Cross is above it on the original bridge which was replaced by the famous "Gridiron" girder structure in 1892. It is passing below Ray Street road bridge. Note the peculiar signal spectacles on the extreme right: perhaps as with the locomotive a degree of artistic licence has crept in here? Likewise the track. Six rails can be seen so the GN train is on the nearest of three tracks but there were only ever two here. Also the nearest would have been used by Down trains going away from the artist's position. The locomotive's exhaust looks somewhat 'feeble' considering that the train would have been climbing the 270 yards of 1 in 40 from the dip under the Met to be level with it at Farringdon Street station.*

*The area where the workmen are gathered is a mystery. It is shown as being lower than the GN's running lines which are on the aforementioned 1 in 40 gradient. Yet from 1873-74 onward when Farringdon Street goods depot was built the land between the retaining wall on the left and the track on which the train is running was considerably higher than is shown here: indeed it was at the same level as the station and accommodated the goods depot's headshunts. A hydraulic pumping station was built here (the cranes and hoists in the depot were hydraulically powered).*

*There is a long single-piece ladder between 'ground' level and the nearest bridge, above the parapet of which appears the jib of a crane with a rope descending to the work area. At the time of the engraving this bridge did not carry a road. A new girder bridge was constructed over the railway in 1875-76 when Clerkenwell Road was built and it is likely that prior to that, this structure carried a sewer – possibly the enclosed Fleet River. There were various works in this area at the time to re-route and enhance the Fleet sewer and its associated feeder and relief channels but while a fascinating subject in itself it is rather irrelevant to the story of the railways.*

*The other two bridges are Vine Street (brick arches on girders) and Ray Street (brick single arch). Beyond them are the mouths of the two Clerkenwell Tunnels, the one to the left being the Met's cut-and-cover which was built first and was covered by Farringdon Road and the lower, later GNR bored one to the right.*

*Fig 14 (left). A 1950s photograph from the London Transport Museum collection (Ref: M114) showing the "Gridiron" looking towards King's Cross with the mouth of the Met's Clerkenwell Tunnel on the left under the arch. The 'patched' holes in the brickwork of the parapet are a mystery but may have been a WW2 measure to enable the fire service to pump water from a sump under the GN's tracks. Similar apertures still exist (and can be used) in parapets of road bridges over the canals in Birmingham.*

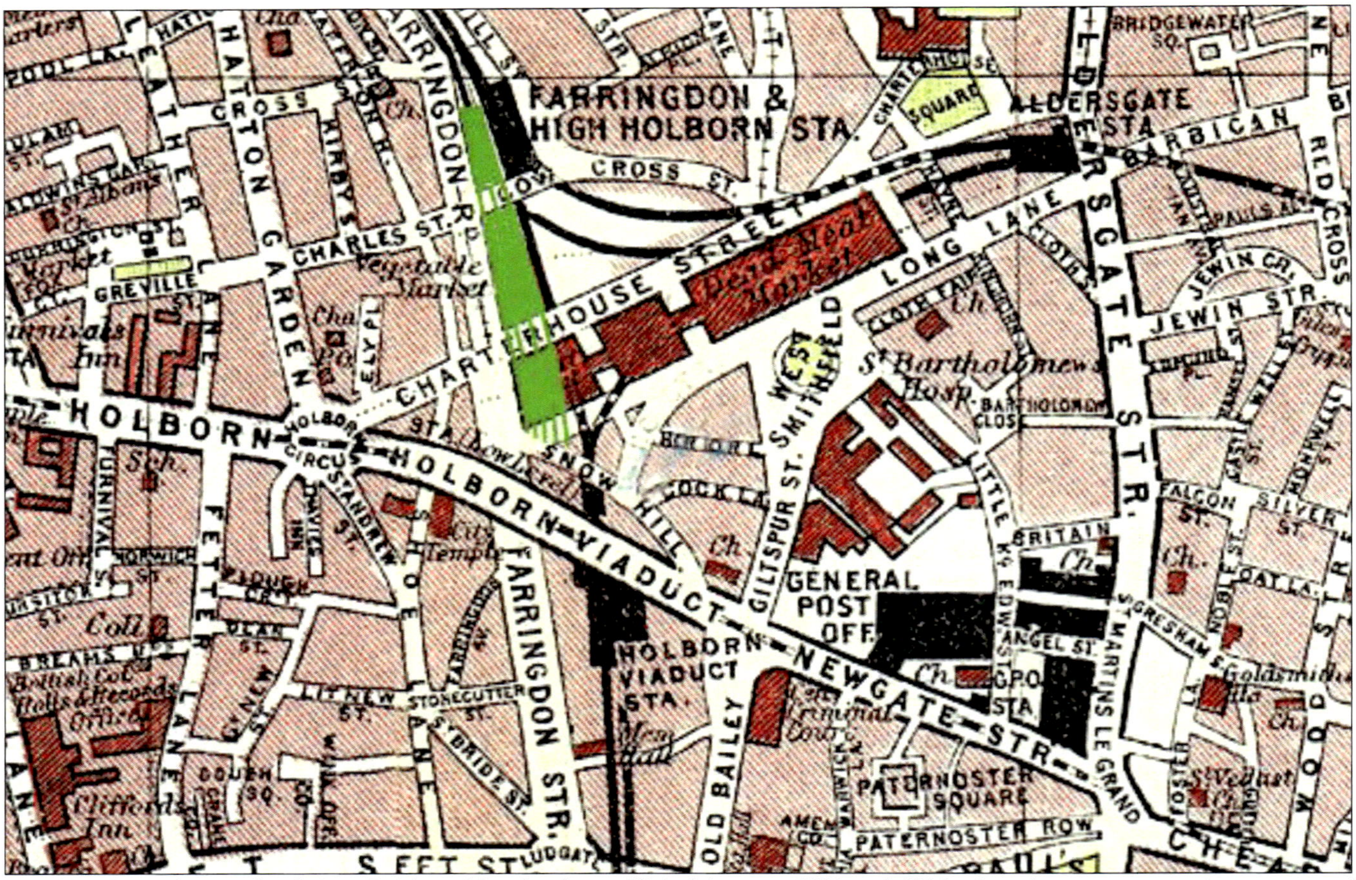

Fig. 15. *Part of an undated A to Z Atlas of London but published between 1922 and 1936 when the passenger station was named "Farringdon & High Holborn". Smithfield's building is identified as "Dead Meat Market". The area occupied by the GNR goods depot has been shaded green to show the extent of its subterranean presence south of Cow Cross Street.*

Fig.16. *The third station building, designed by Charles W Clark and opened in 1923, still stands and will be familiar to many. The exterior is not what it may seem, being white-coloured faience to simulate stonework. It bears the name "Farringdon & High Holborn" by which it was known between 1922 and 1936. Photograph by W H R Godwin in 1957. The car on the extreme right coming out of Turnmill Street is an Armstrong-Siddeley Sapphire 346.*

# 7  Signalling at Farringdon

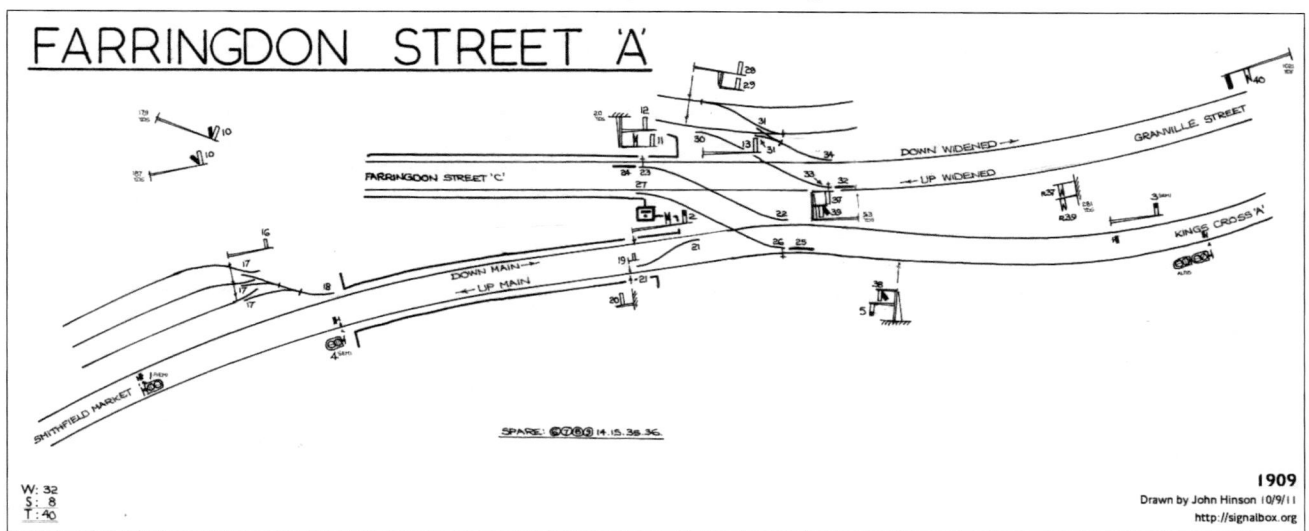

*Fig. 17. Signalling was provided by the Metropolitan Railway and its successors.  Farringdon Street 'A' signal box was on the north end of the westbound/clockwise Metropolitan/Circle platform.  This 1909 diagram shows the connections between the Widened Lines and the electrified Metropolitan/Circle tracks.  Note the double crossovers between the Widened Lines and the depot tracks.  (John Hinson, www.signalbox.org).*

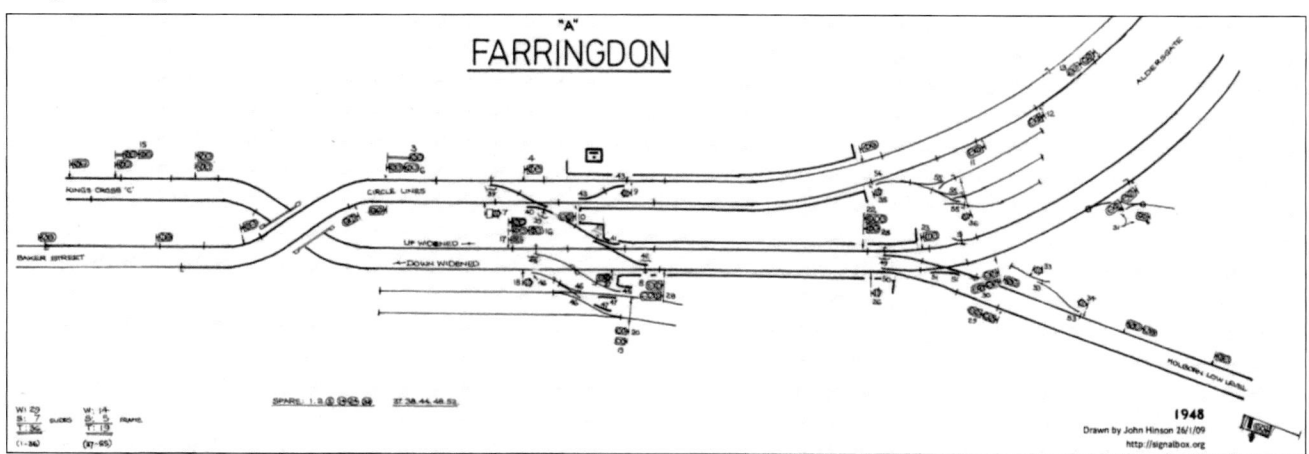

*Fig. 18. Diagram of the new Farringdon 'A' box, 1948, re located on the eastbound/anti-clockwise Met/Circle platform . Semaphore signals have given way to colour lights and there was now only a single track with slips connecting the Metropolitan and Widened Lines. (John Hinson, www.signalbox.org).*

*Fig. 19. Photographs showing Widened Lines signals are rare but this splendid example shows one at Aldersgate Street circa 1921.  It obviously applies to the Down direction but is on the Up (Moorgate-bound) platform.  MR 0-4-4T No. 1530 heads a Down train from the City to Bedfordshire while a Metropolitan electric train waits to go towards Paddington.  The soot-stained glass in the roof prevents the afternoon sunshine getting through to the interior of the station.*

# 8 The Goods Depot

An overall plan of the 1894 enlargement of the depot is shown on pages 24/25. The main building rested on a substructure at railway level comprising piers and arches of Staffordshire blue brick and cast-iron columns with wrought-iron girders running across. In the upper storeys, the floors were supported on cast-iron columns or wrought-iron stanchions and box girders where large open areas were required. The floors themselves saw the first use of Westwood, Baillie & Co's patent corrugated-truss fireproof system which comprised corrugated-iron plates riveted together and bolted to wrought-iron trusses underneath and filled on the upper side with concrete and asphalt. The windows were Moline's patent wrought-iron sashes. The roofs were of also of iron construction. The street frontage employed Pether's pressed bricks with window arches of red brick and Portland stone keys.

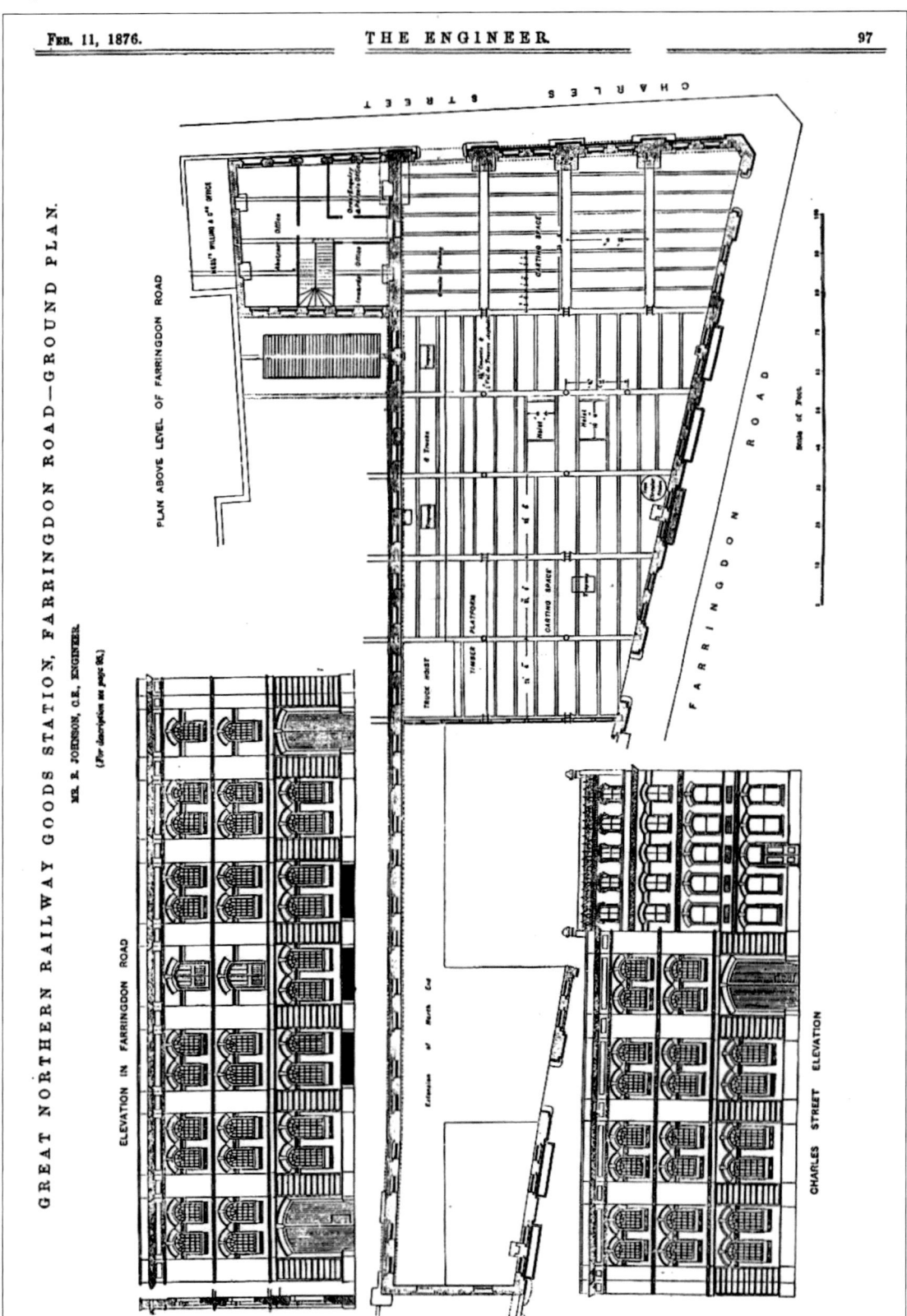

*Fig. 20. Ground plan and west/south elevations of prior to the 1894 enlargement, from THE ENGINEER of 11th Feb 1876 (Adrian Marks).*

Goods-handling machinery included hydraulically powered cranes in the basement (the two largest able to raise three tons each), wagon turntables and traversers, capstans for rope hauling railway wagons and two hoists to move wagons between the basement and first floor. Two goods lifts of two tons capacity each and four jiggers for loads up to 10cwt served all levels of the building.

The south end of the depot was parallel with but not connected to the LC&DR's Smithfield Sidings (also known as Snow Hill Sidings and nothing to do with the Meat Market). These were opened on 19th December 1881 to stable stock of trains using Holborn Viaduct station and unusually one pair of trqcks included a sector table for locomotive release, there being no room for a conventional release crossover or a turntable.

Two of the sidings are still in use today as an emergency 'bolthole' for any Thameslink trains that fail the DC/AC changeover at City Thameslink station.

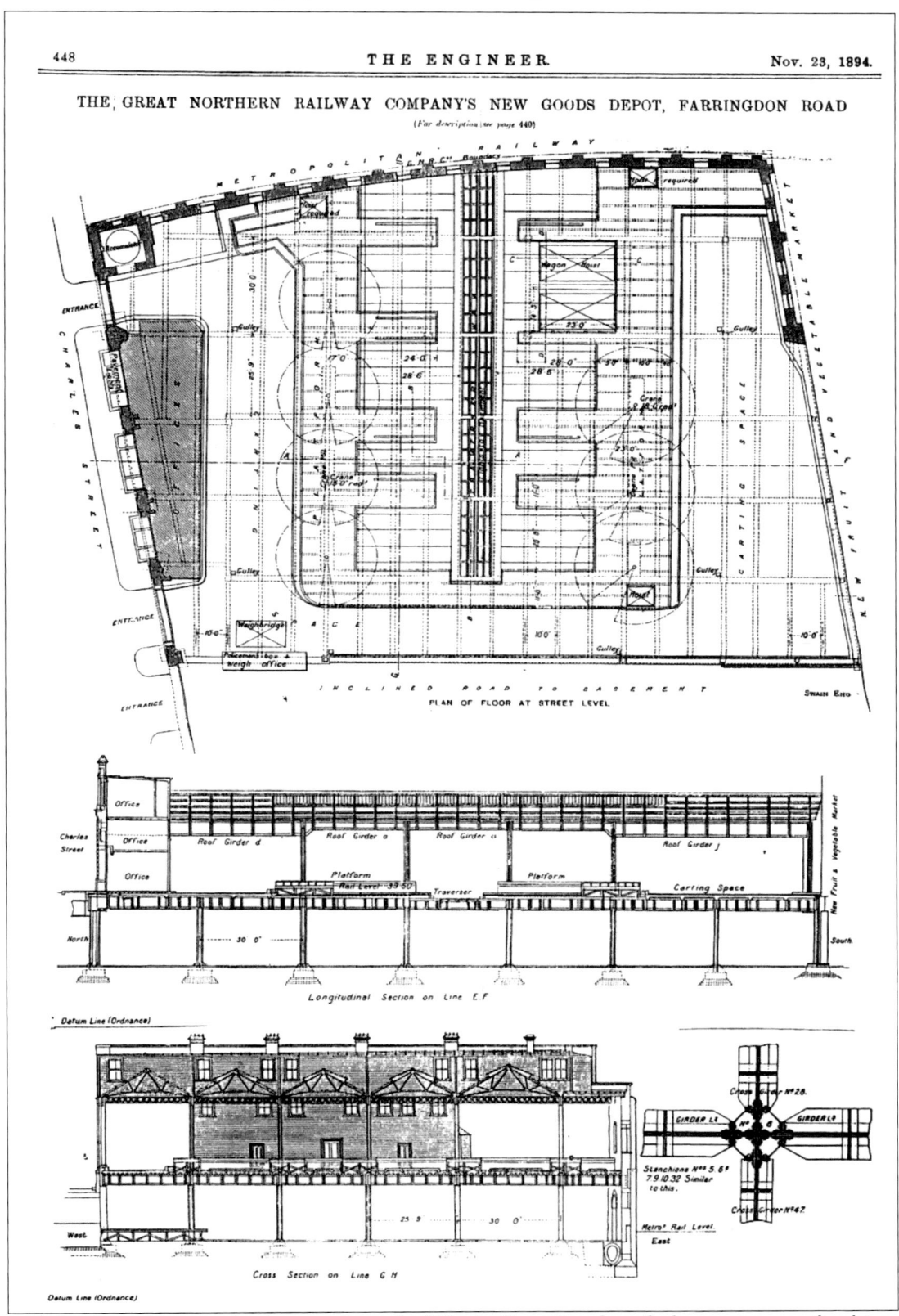

*Fig. 21. Plan and elevations of the street level area to the south of Charles Street (i.e. above the basement area outlined in purple in fig. 28 (page 25). (Adrian Marks).*

Farringdon Street depot seems to have settled into a busy but uneventful life of service. The main traffic would have been inbound foodstuffs for the three markets above it and the nearby Smithfield Meat Market. There were also supplies to and finished products from local industries including printing works, a distillery and a snuff factory.

Although the passenger station was renamed Farringdon & High Holborn on 26th January 1922 and (plain) Farringdon on 21st April 1936, the goods depot kept its original name although as seen on fig. 28, it was shown as Farringdon Road on the GNR's own plan in THE ENGINEER IN 1894. On 1st January 1923 the depot passed into the ownership of the LNER.

The first year of World War Two had little effect on the depot but everything changed on one night in September 1940 when a German high explosive bomb caused damage which prevented trains running in or out for a while, although normal working was resumed fairly quickly. Sources vary as to the exact date. BROOKSBANK has it as the 24th but the GNRS' Collection has a photograph of the damage that was contemporaneously annotated as the 28th. Other photographs in the Collection of war damage elsewhere carry the dates they were taken, usually the same day as or the day after the raid, so perhaps this was not photographed until four days after the event.

Much more serious was the combined effect of 24 hours of raids on the 16th and 17th of April 1941. The offices were destroyed by fire, the main building seriously damaged and its contents destroyed. The basement area was blocked by debris and water. Power supplies, including hydraulic power, failed. Two wagons and twelve [road] cartage vehicles were destroyed. The depot was closed and traffic diverted to King's Cross and Tuffnell Park. The Low Level reopened on 12th May but only for potato traffic. The date of further reopening is not known.

Although the Blitz ended in 1941, there were still air raids

on a less intensive scale and another at 8 am on 24th August 1944 caused further damage. The LNER still referred to it as "Farringdon Street" in its report which includes the rather understated comment "working impeded".

The next report is on 8th March 1945, referring to "Farringdon" (no "Street") and the arrival of a V2 rocket on the market above at 11.03. It penetrated to the depot below causing considerable damage and burying tracks and equipment in debris.

Although partially reopened on 22nd March normal working did not resume until 3rd May.

Farringdon Street goods depot had never been easy to operate and the wartime damage only made things worse. Greater use was being made of road motor transport (notably Scammell "mechanical horses" and articulated semi-trailers) which meant that traffic to and from the Farringdon area could be more easily handled by King's Cross goods depot. Finally, as noted in the introduction, the overall volume of traffic was diminishing as manufacturing industry declined.

Thus it was that the depot closed with effect from 16th January 1956. The Cowcross Street (Charles Street) frontage, which bore a marked similarity to the front of company's Royal Mint Street goods depot although the brickwork was less ornate, is shown in fig. 22 below. The 'Policeman's Box and Weigh Office' shown at bottom left of the plan in Fig. 21 (page 19) are at the extreme right in the photograph. This part of the depot is thought to have been demolished shortly after closure but despite increasing internal dereliction the main building alongside the passenger station remained substantially intact for a further 32 years, as will be seen in the photographs in chapter 10. The trackless ground was used, as were many London "bomb sites" and other vacant plots in the 1960s and 70s, as a NCP car park, reached by a precarious-looking ramp from Farringdon Road via a 'hole in the wall' at the north end of the main building.

Fig. 22. The remains of the surface level building on the south side of Cowcross Street. It is 1954 or later: the middle of the three cars is a Hillman Husky, introduced in that year. The words "Goods Depot" can just be discerned above the nearest two arches. The middle poster appears to extol the virtues of "Beer" without naming a brand, the next advertises gas using the once familiar logo "Mr Therm" created for the Gas Light & Coke Co in 1931 and adopted by the nationalised gas boards in 1948. Finally, still very much with us, an advert for Mackintosh's 'Quality Street' sweets. The "Policeman's box and weigh office" shown on maps are behind the gate marked "L N E R GOODS DEPOT" which is at the top of the ramp down to basement level. The (road vehicle) weighbridge shown on the plan is behind the new brick wall in the nearest arch. (National Railway Museum).

# 9  Farringdon From The Air in 1947

In July 1947 Aerofilms Ltd took a series of aerial photographs of Farringdon station. Whether these were commissioned by the LNER or another organisation (the City of London possibly?) is not known. The images are now available as high-resolution scans from the "Britain From Above" website.

*Fig.23. This is the view from the east-southeast, with Ray Street and the 'gridiron' at the top right. The 'black hole' of the road vehicle entrance to the former Vine Street goods depot is prominent at the east (right-hand) end of Vine Street bridge. A J50 0-6-0ST is shunting wagons in the depot and another (unidentifiable) locomotive passes on an Up goods train. The purpose of the canopy over one track and a capstan adjacent to the retaining wall is unknown. The cramped nature of the depot and in particular the shortness of the headshunts is very noticeable. The damaged roof of the warehouse building has been repaired but the offices between the warehouse and the passenger station have been left open.  Facing them across Cow Cross Street is the frontage of the destroyed building seen in the previous photograph.  The adjacent buildings on the overbridge deserve a mention – pity the poor occupants of the soot-stained premises over the steam-worked lines!*

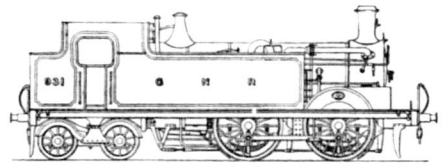

*Fig. 24. The distinctive shape of the warehouse building is seen to good effect in this view looking approximately due north. Farringdon Road runs diagonally from top left to bottom right, passing under Holborn Viaduct. The subterranean tracks of the depot ran as far as Snow Hill, marked on the photograph.*

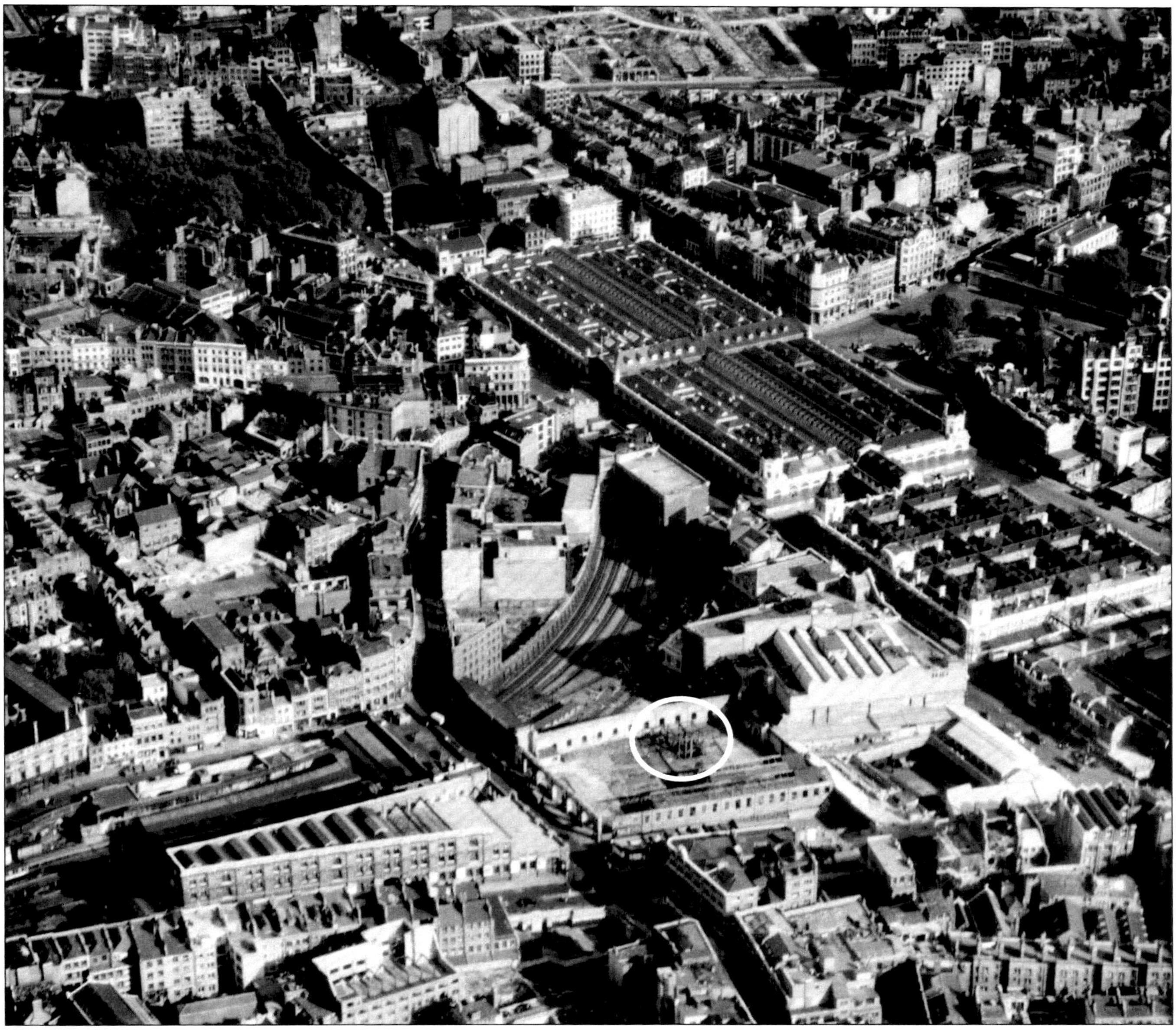

*Fig. 25.   Looking almost due east, Farringdon Road runs north-south alongside the goods station towards the bottom of the picture. Prominent in the centre is the flattened area and shell of the buildings that once held the markets above the subterranean depot, with the metal structures of the wagon hoists visible (circled, and enlarged in Fig. 26 [below]).  The size of Smithfield meat market can be seen to good effect in this view and beyond it, the cavernous roof of Aldersgate Street station.*

*Fig. 26.*

23

# Farringdon Passenger Station and Goods Depot 29th August 1952

*Fig.27. This view looking south from Clerkenwell Road bridge shows to good effect the tapering width from south to north of the warehouse, built on the site of the first passenger station. On the extreme left behind the parapet of the eastern retaining wall is a glimpse of the GNR's Clerkenwell or Turnmill Street Stables, featured in Chapter 13. Below the parapet the retaining wall has a substantial plate girder over the 'gap' originally occupied by the short-lived turntable. Next are the Metropolitan electric lines with a Westbound train of 'T' stock, then the Widened Lines with an N2 and GN-line train at the Down platform. The sign on the end of the main building proclaims "LONDON & NORTH EASTERN RAILWAY CITY GOODS STATION". The purpose of the canopy over the right-hand headshunt is unclear. There is no sign of a wheeldrop or lifting apparatus here for repairs to wagons such as would warrant a covered hoist to lift one end of a wagon to remove a wheelset. The position of the capstan in the middle of the space between the track and the canopy support would prevent a road vehicle standing alongside for transfer of a weather-sensitive consignment. Both the canopy and capstan are marked on the plan below but with no description for the former. Farringdon Road is on the extreme right, behind parapet of the western retaining wall. (Frank Goudie).*

# GNR Basement level plan, from THE ENGINEER, 23rd November 1894

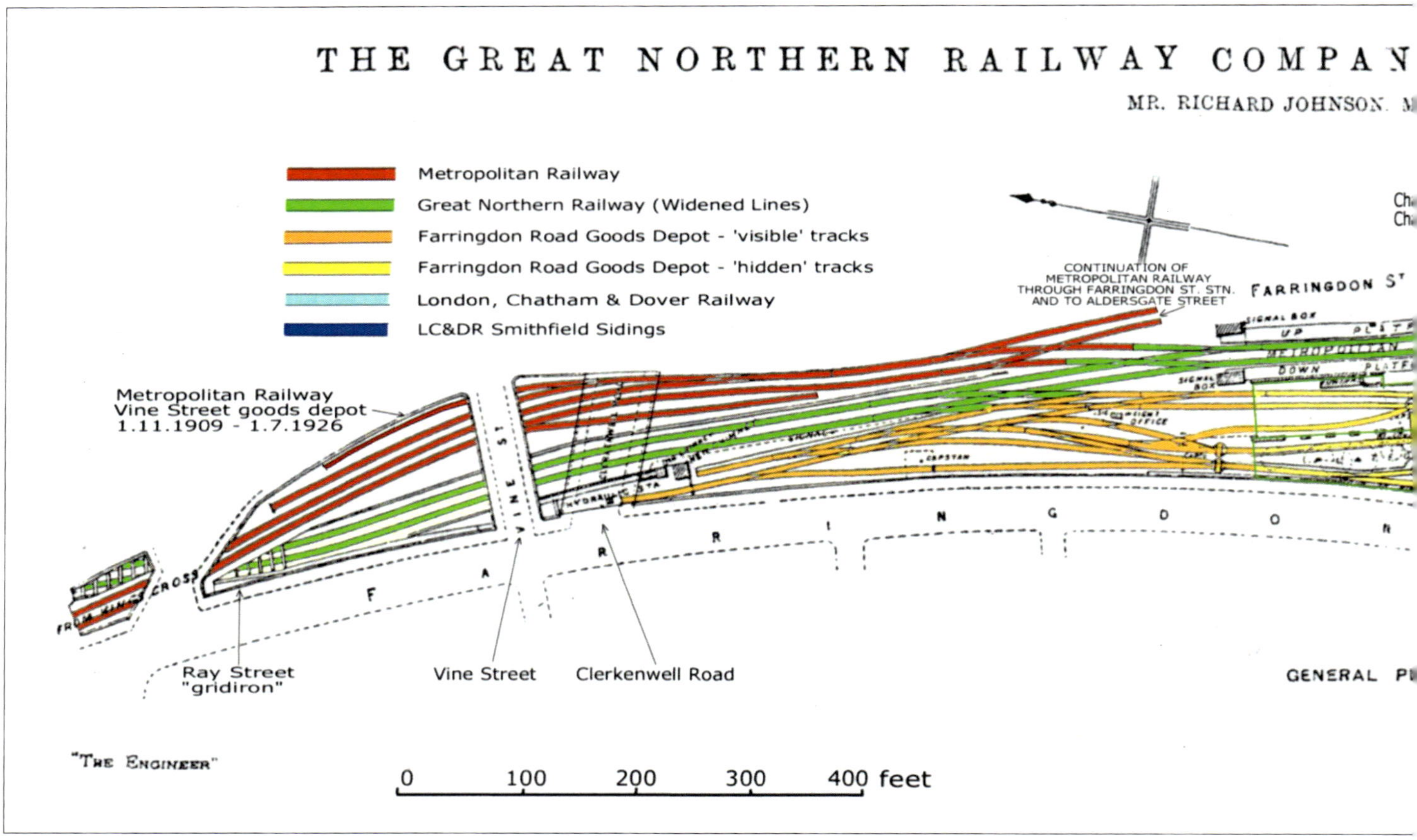

THE GREAT NORTHERN RAILWAY COMPAN

MR. RICHARD JOHNSON. M

- Metropolitan Railway
- Great Northern Railway (Widened Lines)
- Farringdon Road Goods Depot - 'visible' tracks
- Farringdon Road Goods Depot - 'hidden' tracks
- London, Chatham & Dover Railway
- LC&DR Smithfield Sidings

CONTINUATION OF METROPOLITAN RAILWAY THROUGH FARRINGDON ST. STN. AND TO ALDERSGATE STREET

FARRINGDON St

Metropolitan Railway Vine Street goods depot 1.11.1909 - 1.7.1926

Ray Street "gridiron"    Vine Street    Clerkenwell Road

GENERAL PI

"THE ENGINEER"

0    100    200    300    400 feet

*Fig. 28. This plan as originally published was of course in monochrome but the various lines are shown above coloured to indicate their descriptions and owning companies. The various significant features have also been annotated and Vine Street goods depot, which did not exist in 1894, has been added. As the plan was primarily intended to illustrate the GNR's goods depot the continuation of the Metropolitan Railway tracks [red] to Aldersgate Street was not shown on the original.*

*The plan shows individual items of handling equipment although it is difficult to distinguish capstan 'dots' from unintended marks. However, it can be seen that there were 25 wagon turntables, 13 (possibly 14) rotating cranes and two wagon hoists in the 'subterranean' part of the depot plus a further two turntables outside. The use of a sector table for locomotive release was rare on main line railways in the UK but one existed here, at the end of the westernmost of the LC&DR's Smithfield Sidings. No doubt inches of usable siding*

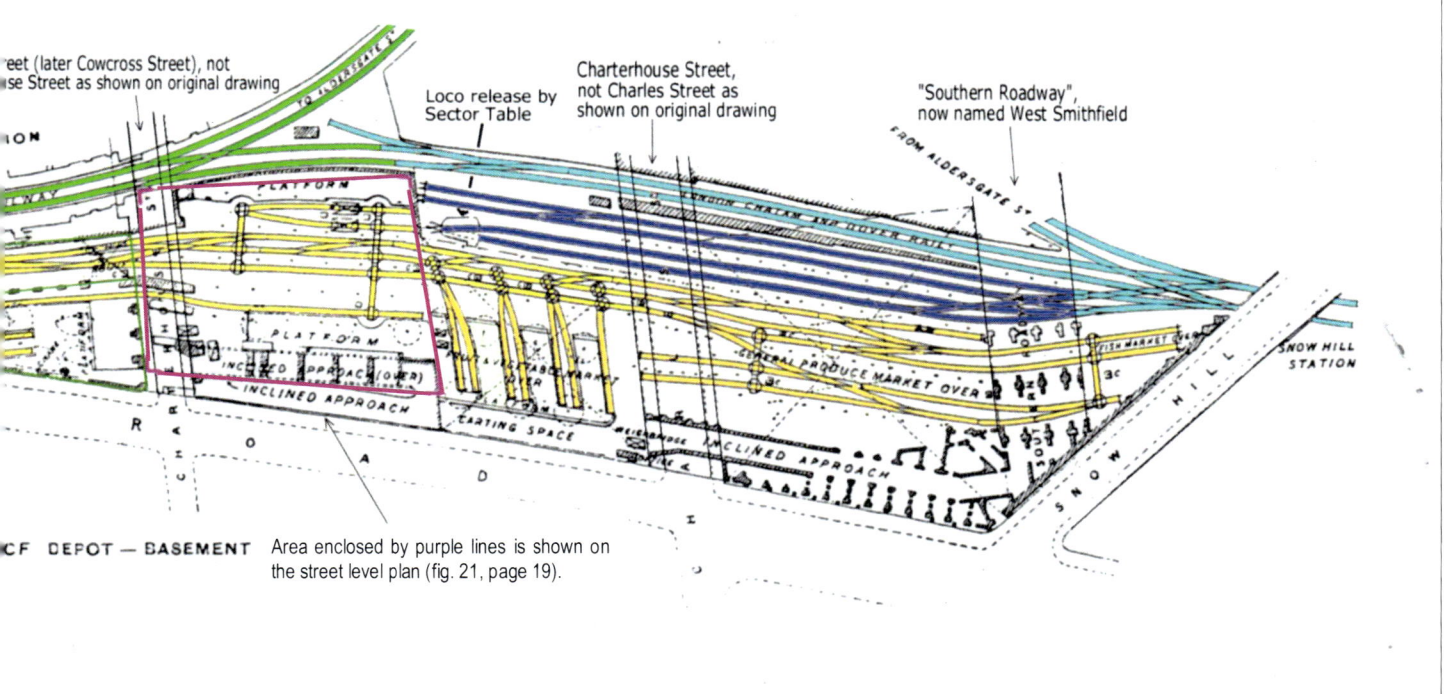

## S NEW GOODS DEPOT, FARRINGDON ROAD

ST. C.E., CHIEF ENGINEER

eet (later Cowcross Street), not
se Street as shown on original drawing

Loco release by
Sector Table

Charterhouse Street,
not Charles Street as
shown on original drawing

"Southern Roadway",
now named West Smithfield

Area enclosed by purple lines is shown on
the street level plan (fig. 21, page 19).

length were critical in this confined location so a conventional crossover could not be used, nor even a turntable. Note too the spur for the banking locomotive for trains requiring assistance up the gradient to Ludgate Hill and Blackfriars. The covered markets are shown: Fruit & Vegetable Market (north [left] of Charterhouse Street), General Produce Market (between Charterhouse Street and West Smithfield) and Fish Market (between West Smithfield and Snow Hill).

There is a significant error on the part of the draughtsman: he has transposed the names of Charles Street and Charterhouse Street. These have been corrected and the modern day name "West Smithfield" added for the road shown in 1894 as "Southern Roadway". In addition he committed the minor error of "Chatam" instead of "Chatham" in the company's title.

(Scan sourced by Adrian Marks)

# 10 The Goods Depot After Closure

From early 1956 the depot went into a 32-year period of "benign neglect". There was no apparent urgency on British Railways' part to sell the site for development but this allowed a few intrepid industrial archeologists to explore the more remote parts up to the 1970s when National Car Parks became the tenant.

*Fig. 29. At the junction of Farringdon Road and Charles Street looking north-east. The most severe bomb damage appears to be undergoing stabilisation (it was never fully repaired). Trolleybus wires are still in place so this photograph was taken no later than January 1962. (National Railway Museum ref: Liverpool Street/1995-7233_LIVST_DF_250)*

*Fig. 30. A 1980s view from the same point. The passenger station is the light coloured building in Cowcross Street on the extreme right. (Jonathan David)*

*Fig.31. The ragged edge of the damaged brickwork had been made a little neater by forming a slope. The top floor had lost some of the window frames. Note how the lighter Portland Stone keystones stand out against the arches. (Jonathan David).*

*Fig. 32. The northern end of the Farringdon Road elevation: despite the ravages of war, disuse and pollution it was still an impressive structure. (Jonathan David).*

27

*Fig. 33 (left): The northernmost doorway in the Farringdon Road elevation, with faded "L . N . E . R GOODS DEPOT" just discernible.*
*Fig. 34 (right): The eastern doorway in the Cowcross Street elevation. The presence of the Ford Escort Mk.III Estate dates this as no earlier than September 1980. (both: Jonathan David)*

*Fig. 35. An undated view showing the roofless remains of the canopy in situ between the depot building and the wall of the passenger station. The tracks in stone setts disappear under Cowcross Street to the rest of the depot. Note also the capstans for rope shunting (using horses). The absence of soot on the right-hand end of the valancing indicates that the track nearest the depot building, was not normally used by steam locomotives. (Adrian Marks)*

28

*Fig. 36. The pointwork under Charterhouse Street and the General Produce Market. It looks as if even this section was latterly used as a car park, hence the arrows painted on the columns, the warning stripes on the downpipe and the fluorescent lighting. Smithfield Sidings are behind the columns on the left. (Nick Catford / Disused Stations)*

*Fig. 37: Looking south from the car park ramp over the site of the former canopy to the remains of the offices built over the tracks going under Cowcross Street. Note the tree!*

*Fig. 38: Looking north from the remarkably well-established tree in the former offices.*
*(both: Jonathan David).*

*Fig. 39 (above): The view from Turnmill Street on the east side of the railway cutting. Note the different advertisements on the end of the building. Unsurprisingly there are several Ford Escorts and Cortinas in the NCP car park plus a Mini, a VW Beetle and a Triumph Dolomite. ('Farringdon Road', Survey of London Vol. 46.)*

*Fig. 40 (right): A telephoto view from Clerkenwell Road bridge, again showing the shallow concave curve of the Farringdon Road elevation and the taper outwards from north to south. (Jonathan David).*

*Fig. 41. Looking south along Farringdon Road. The open gateway leads to the ramp down to the car park. 'Fags and beer' adverts, as on the end wall, were once commonplace but seem alien to us now. How times, tastes and legislation have changed! (Jonathan David).*

*Fig. 42. The replacement office block, built by Bovis Construction in 1988-92 to the design of the Siefert Group, which won the architectural critic Hugh Pearman's verdict of "one of the two worst buildings of 1992" and his description of the style as "Early Learning Centre architecture"! (Allan Sibley, 12th December 2018)*

# 11  Goods Traffic at Farringdon Street

Photographs of the depot in its working days are rare – those contained in this book are all from the post-1945 period when wartime damage had rendered a large part of the premises unusable.  But activity there certainly was because Table 1 (data extracted from the WTT of July 1897) shows that only three years after its enlargement, the depot saw between 21 and 26 trains arriving each weekday, 10 to 15 of them between midnight and 8am, after which goods trains were prohibited from the Widened Lines for a couple of hours to cater for the morning passenger 'peak' service.  There was, of course, an equivalent number of outbound trains and between five and seven each way on Sundays.

Maximum loads varied on the through route to the LC&DR depending on the locomotive type, whether full wagons or empties and other factors (See Table Fig. 46, page 37).

However the maximum train length to and from Farringdon Street goods depot was always 28 wagons.  A speed restriction of 6 mph applied entering the depot and 25 mph leaving.

Through workings from and to distant places were commonplace, for example in 1935 the 7.29pm departure to King's Cross goods conveyed wagons to be transferred to the 8.30pm to Manchester Deansgate and the 9pm to Liverpool Huskisson.  Other later trains had wagons for Leeds, Nottingham and York.

Table 2 shows the services operated between 17th July and 10th September 1933.

Table 3 shows that traffic was still buoyant in 1938 but Tables 4 and 5 reveal the effects of the damage caused during the Second World War.

---

**Table 1:  Arrivals at and departures from Farringdon Street Goods Depot, July 1897.**
Data extracted by Paul Taylor from the GNR Working Time Table (Main Line and Suburban Branches) dated 1st July 1897

### Arrivals

| No. | Time | From | Runs | Notes |
|---|---|---|---|---|
| | | Monday - Saturday | | |
| 4 | 12 47am | East Goods | | a 11 05pm |
| 13 | 1 27 | East Goods | MX R | a 12 25am |
| 19 | 2.22 | East Goods | | |
| 26 | 3 7 | East Goods | | |
| 33 | 3 52 | East Goods | R | b 3 25 |
| 37 | 4 37 | East Goods | | b 4 10 |
| 46 | 5 12 | East Goods | R | b 4 45 |
| 50 | 5 41 | East Goods | | b 5 15 |
| 57 | 6 0 | East Goods | R | a 5 10  b 5 32 |
| 60 | 6 12 | East Goods | | a 5 25  b 5 45 |
| 100 | 6 38 | East Goods | | b 6 10 |
| 104 | 6 58 | King's + Gds | R | Met eng and bk |
| 105 | 7 5 | East Goods | | b 6 39 |
| 108 | 7 18 | East Goods | | b 6 52 |
| 119 | 7 58 | East Goods | | b 7 30 |
| 224 | 10 48 | East Goods | | a 9 40 |
| 286 | 12 34pm | East Goods | SX | a11 45 |
| 375 | 3 3 | King's + Gds | | |
| 419 | 3 57 | East Goods | | a 2 58pm |
| 532 | 7 0 | East Goods | | |
| 567 | 7 54 | East Goods | | |
| 582 | 8 14 | East Goods | | |
| 615 | 9 15 | East Goods | | a 8 25 |
| 634 | 9 50 | King's + Gds | | |
| 653 | 10 49 | East Goods | | a 9 50 |
| 675 | 11 37 | East Goods | SX | a10 35 |
| 683 | 11 55 | Kings + Gds | SO | |
| | | | | |
| | | Sunday | | |
| 8 | 1 27am | East Goods | | |
| 17 | 2 55 | East Goods | R | |
| 26 | 4 40 | East Goods | R | |
| 33 | 5 40 | East Goods | | a 4 50am |
| 55 | 6 56 | East Goods | | a 6 20 |
| 160 | 11 12pm | East Goods | | a10 30pm |
| 165 | 11 43 | East Goods | | a10 50 |

### Departures

| No. | Time | To | Runs | Notes |
|---|---|---|---|---|
| | | Monday - Saturday | | |
| 25 | 12 15am | Clarence Yard | | K+S |
| 41 | 1 16 | Clarence Yard | | K+S |
| 58 | 2 0 | Clarence Yard | | K+S |
| 68 | 2 38 | Clarence Yard | | |
| 76 | 3 0 | Ferme Park | MX R | |
| 84 | 3 37 | Clarence Yard | MX R | |
| 96 | 4 15 | Clarence Yard | R | |
| 113 | 5 4 | Clarence Yard | | |
| 132 | 5 25 | Clarence Yard | R | |
| 149 | 6 1 | King's + Gds | R | Met Ety Wgns |
| 156 | 6 18 | Clarence Yard | | |
| 163 | 6 30 | Clarence Yard | | |
| 181 | 7 1 | Clarence Yard | | |
| 207 | 7 39 | King's + Gds | | |
| 221 | 7 51 | King's + Gds | | |
| 307 | 10 32 | Clarence Yard | | |
| 328 | 11 34 | King's + Gds | | |
| 394 | 1 49pm | Clarence Yard | SX | |
| 481 | 3 36 | Clarence Yard | | |
| 582 | 6 35 | Clarence Yard | | |
| 611 | 7 26 | Clarence Yard | | |
| 643 | 8 31 | Clarence Yard | | |
| 665 | 9 12 | King's + Gds | | |
| 689 | 10 5 | Clarence Yard | | |
| 702 | 10 31 | King's + Gds | | |
| 711 | 11 15 | King's + Gds | | |
| | | | | |
| | | Sunday | | |
| 14 | 12 15am | Clarence Yard | | |
| 37 | 1 16 | Clarence Yard | | |
| 47 | 2 0 | Clarence Yard | | |
| 60 | 3 20 | Clarence Yard | R | |
| 71 | 5 2 | Clarence Yard | R | |
| 77 | 6 1 | Clarence Yard | | |
| 88 | 7 25 | King's + Gds | | |

East Goods is at Finsbury Park

a = Engine and brake to leave King's Cross Goods Yard  at
time stated

b = Starts from Hornsey Sidings, when required, at time
stated and will not stop at East Goods Yard

Met eng and bk = Metropolitan Railway engine and brake

M = Monday    S = Saturday
O = only       X = excepted    R = runs if required

Clarence Yard is at Finsbury Park

Ferme Park is between Harringay and Hornsey

K+S = To stop at King's Cross Suburban station when necessary for Farringdon Street staff to alight.

Met Ety Wgns = Metropolitan Railway empty wagons to be worked by the Metropolitan Company.

**Table 2: Arrivals at and departures from Farringdon Street Goods Depot, 17th July to 10th September 1933.**
Data extracted by Allan Sibley from the LNER Working Time Table [London (Suburban), Hatfield, Hitchin and Branches] supplied by Steve White.

## Arrivals

| No. | Time | From | Runs | Notes |
|---|---|---|---|---|
| | Monday - Saturday | | | |
| 6 | 12 47am | King's + Gds | MX | |
| 10 | 12 58 | East Goods | MO | |
| 15 | 1 27 | Ferme Park | MX R | |
| 28 | 2 22 | Ferme Park | MX | East Goods MO |
| 37 | 3 7 | Ferme Park | MX | East Goods MO |
| 49 | 3 50 | Ferme Park | MX R | East Goods MO R |
| 50 | 4 37 | Ferme Park | MX | East Goods MO |
| 57 | 5 2 | Ferme Park | MX | |
| 63 | 5 12 | Ferme Park | MX | East Goods MO |
| 65 | 5 41 | Ferme Park | MX | East Goods MO |
| 75 | 6 5 | Ferme Park | MX R | East Goods MO R |
| 85 | 6 16 | Ferme Park | MX | East Goods MO |
| 121 | 6 38 | Ferme Park | MX | East Goods MO |
| 130 | 7 9 | Ferme Park | MX | East Goods MO |
| 133 | 7 16 | Ferme Park | MX | East Goods MO |
| 148 | 8 11 | Ferme Park | MX | East Goods MO |
| 256 | 10 6 | York Road | | A |
| 287 | 10 45 | Holloway | R | B |
| 366 | 12 37pm | East Goods | SX | |
| 483 | 3 18 | East Goods | SX | |
| 527 | 3 47 | East Goods | SO R | |
| 531 | 4 2 | East Goods | SO | C |
| 558a | 4 46 | East Goods | SO R | |
| 573 | 4 59 | East Goods | SO R | |
| 665a | 6 9 | Moorgate St | SO | LE D |
| 642 | 6 35 | East goods | | |
| 699a | 7 46 | East Goods | | E |
| 724 | 8 15 | East Goods | R | |
| 752 | 8 22 | Moorgate St | SX | F |
| 738 | 8 37 | East Goods | SO R | |
| 764 | 9 9 | East Goods | | |
| 793 | 9 46 | King's + Gds | | |
| 819 | 10 49 | Clarence Yard | | Engine & Brake |
| 845 | 11 37 | East Goods | | |
| 856 | 12 6am | East Goods | SX | Meat |
| | | | | |
| | Sunday | | | |
| 10 | 1 27am | East Goods | | |
| 16 | 2 55 | East Goods | R | |
| 21 | 5 40 | East Goods | | |
| 28 | 6 56 | East Goods | | |
| 187 | 9 29pm | King's + Gds | R | Engine & Brake |
| 218 | 12 3am | East Goods | | Meat |

## Departures

| No. | Time | To | Runs | Notes |
|---|---|---|---|---|
| | Monday - Saturday | | | |
| 24 | 12 14am | Clarence Yard | MX | |
| 44 | 12 56 | Clarence Yard | MO | |
| 46 | 12 59 | Clarence Yard | MX | |
| 47 | 1 15 | Clarence Yard | MO | |
| 57 | 2 0 | Clarence Yard | | |
| 68 | 2 32 | Clarence Yard | | |
| 77 | 3 0 | Ferme Park | MX R | |
| 85 | 3 40 | Clarence Yard | MX R | |
| 96 | 4 15 | Clarence Yard | R | |
| 119 | 5 4 | Clarence Yard | | |
| 154 | 6 22 | Clarence Yard | | |
| 162 | 6 34 | Clarence Yard | | |
| 187 | 7 5 | Clarence Yard | | |
| 205 | 7 34 | Clarence Yard | | |
| 227 | 7 55 | Clarence Yard | | |
| 242 | 8 17 | Clarence Yard | R | Light Engine |
| 352 | 10 38 | Clarence Yard | | |
| 375 | 11 36 | Clarence Yard | | |
| 469 | 1 49pm | Clarence Yard | SX | G |
| 563 | 3 41 | Clarence Yard | SX | |
| 583 | 4 15 | Clarence Yard | SO R | |
| 593 | 4 33 | Clarence Yard | SO R | |
| 608 | 4 59 | Clarence Yard | SO R | |
| 618 | 5 12 | Clarence Yard | SO R | |
| 685 | 6 35 | King's + Gds | SO | |
| " | 6 39 | King's + Gds | SX | |
| 719 | 7 29 | King's + Gds | | H |
| 759 | 8 31 | King's + Gds | SX | Clarence Yd SO |
| 769 | 8 50 | King's + Gds | | |
| 783 | 9 9 | Clarence Yard | SX | |
| " | 9 12 | King's + Gds | SO R | J |
| 793 | 9 25 | King's + Gds | SX | |
| 801 | 9 45 | Clarence Yard | SO | |
| 811 | 9 55 | King's + Gds | SX | |
| 821 | 10 27 | Clarence Yard | SX | |
| " | 10 31 | Clarence Yard | SO | |
| 851 | 11 14 | Clarence Yard | | |
| | | | | |
| | Sunday | | | |
| 14 | 12 16am | Clarence Yard | | |
| 37 | 1 6 | Clarence Yard | | |
| 44 | 2 0 | Clarence Yard | | |
| 55 | 3 20 | Clarence Yard | R | |
| 61 | 6 1 | Clarence Yard | | |
| 68 | 7 25 | Clarence Yard | | |
| 218 | 9 50pm | King's + Gds | R | K |

East Goods is at Finsbury Park

M = Monday    S = Saturday
O = only    X = excepted    R = runs if required

A  Light engine dep York Road 9 59 am, banking engine for South London trains [to banking engine siding].

B  Farringdon Street to advise Control and East Goods Yard when required.

C  Engine and brakes dep King's + Goods 2 58 pm.

D  Light engine dep Moorgate Street 6 5 pm, works 685 down goods.

E  Convey additional brake van, Saturdays excepted, to return with 769 down.

F  Light engine dep Moorgate Street 8 18 pm, works 769 down goods.

Clarence Yard is at Finsbury Park
Ferme Park is between Harringay and Hornsey

G  Arr King's + Gds 2 3 pm, detach wagons and run forward Engine & Brake.

H  Conveys all traffic for connection 773, 774 and 775 down.

J  Load not exceeding 19 wagons and brake. When not more than 16 wagons and brake worked into dead end at entrance to goods yard and backed into yard. When load exceeds 16 wagons worked into yard via goods tunnel. Farringdon Street Goods to telephone Control and Five Arch number of wagons and Control to transmit to Goods and Mineral Junction.

K  Empties

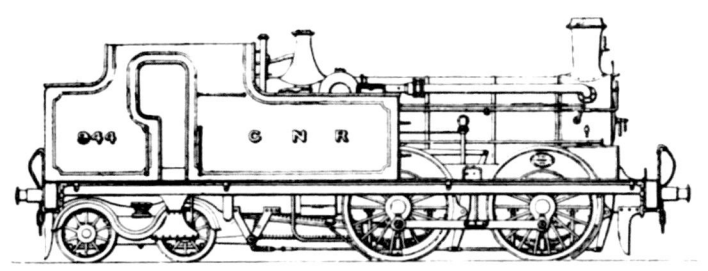

## Table 3: Arrivals at and departures from Farringdon Street Goods Depot, 1938.

Data extracted by Dave Cockle from the LNER Working Time Table (Main Line and Suburban Branches)

### Arrivals

| No. | Time | From | Runs | Notes |
|---|---|---|---|---|
| | Monday - Saturday | | | |
| 6 | 12 47am | Ferme Park | MX | |
| 10 | 12 58 | East Goods | MO | |
| 15 | 1 27 | Ferme Park | MX R | |
| 28 | 2 22 | Ferme Park | | |
| 37 | 3 7 | Ferme Park | | |
| 49 | 3 50 | Ferme Park | R | |
| 50 | 4 37 | Ferme Park | | |
| 57 | 5 3 | Ferme Park | MX | |
| 63 | 5 12 | Ferme Park | | |
| 65 | 5 41 | Ferme Park | | |
| 75 | 6 3 | Ferme Park | R | |
| 85 | 6 16 | Ferme Park | | |
| 116 | 6 38 | Ferme Park | | |
| 123 | 7 9 | Ferme Park | | |
| 140 | 8 11 | Ferme Park | | |
| 287 | 10 45 | East Goods | R | |
| 266 | 12 37pm | East Goods | SX | |
| 483 | 3 18 | East Goods | SX | |
| 525 | 3 47 | East Goods | SO R | |
| 529 | 4 2 | East Goods | SO | |
| 558 | 4 47 | East Goods | SO R | |
| 573 | 4 59 | East Goods | SO R | |
| 642 | 6 35 | East Goods | | |
| 699 | 7 46 | East Goods | | |
| 724 | 8 15 | East Goods | R | |
| 738 | 8 37 | East Goods | SO R | |
| 764 | 9 9 | East Goods | | |
| 793 | 9 46 | King's + Gds | | |
| 819 | 10 49 | Clarence Yard | | |
| 845 | 11 37 | East Goods | SX | |
| | Sunday | | | |
| 10 | 1 27am | East Goods | | |
| 16 | 2 55 | East Goods | R | |
| 21 | 5 40 | East Goods | | |
| 28a | 6 56 | East Goods | | |
| 218 | 12 3am | East Goods | MO | (Monday am) |

| No. | Time | To | Runs | Notes |
|---|---|---|---|---|
| | Monday - Saturday | | | |
| 25 | 12 10am | Clarence Yard | MX | |
| 44 | 12 56 | Clarence Yard | MO | |
| 46 | 12 59 | Clarence Yard | MX | |
| 47 | 1 15 | Clarence Yard | MO | |
| 57 | 1 57 | Clarence Yard | | |
| 68 | 2 32 | Clarence Yard | | |
| 77 | 3 0 | Ferme Park | MX R | |
| 85 | 3 40 | Clarence Yard | MX R | |
| 94 | 4 15 | Clarence Yard | R | |
| 103 | 5 4 | Clarence Yard | | |
| 144 | 6 22 | Clarence Yard | | |
| 153 | 6 34 | Clarence Yard | | |
| 182 | 7 5 | Clarence Yard | | |
| 231 | 7 55 | King's + Gds | | |
| 346 | 10 38 | Clarence Yard | | |
| 375 | 11 38 | King's + Gds | | |
| 469 | 1 49pm | Clarence Yard | SX | |
| 563 | 3 41 | Clarence Yard | SX | |
| 583 | 4 15 | Clarence Yard | SO R | |
| 609 | 4 59 | Clarence Yard | SO R | |
| 620 | 5 12 | Clarence Yard | SO R | |
| 719 | 7 29 | King's + Gds | | |
| 759 | 8 28 | Clarence Yard | | |
| 769 | 8 50 | King's + Gds | SX | |
| 783 | 9 9 | Clarence Yard | | |
| 783 | 9 12 | King's + Gds | SO R | |
| 793 | 9 25 | King's + Gds | SX | |
| 801 | 9 45 | Clarence Yard | SO | |
| 811 | 9 55 | King's + Gds | SX | |
| 821 | 10 25 | Clarence Yard | SX | |
| 821 | 10 31 | Clarence Yard | SO | |
| 851 | 11 14 | Clarence Yard | | |
| | Sunday | | | |
| 14 | 12 16am | Clarence Yard | | |
| 27 | 1 6 | Clarence Yard | | |
| 44 | 2 0 | Clarence Yard | | |
| 55 | 3 20 | Clarence Yard | R | |
| 61 | 6 1 | King's + Gds | | |
| 68 | 7 25 | King's + Gds | | |
| 218 | 9 50pm | King's + Gds | R | |

East Goods Yard is at Finsbury Park

Clarence Yard is at Finsbury Park
Ferme Park is between Harringay and Hornsey

M = Monday    S = Saturday
O = only    X = excepted    R = runs if required

Ferme Park is between Harringay and Hornsey

## Departures

### Table 4: Arrivals At and Departures From Farringdon Street Goods Depot, 1946

Data extracted by Dave Cockle from the LNER Working Time Table (Main Line and Suburban Branches)

#### Arrivals Mon-Sat only

| No. | Time | From | Runs | Notes |
|---|---|---|---|---|
| 1 | 2 28am | East Goods | MX | |
| 2 | 5 38 | Ferme Park | MX | |
| 2 | 5 38 | East Goods | MO | |
| 6 | 10 45 | Ferme Park | | |

#### Departures Mon-Sat only

| No. | Time | To | Runs | Notes |
|---|---|---|---|---|
| 1 | 3 28am | Clarence Yard | MX | |
| 2 | 6 45 | Clarence Yard | | |
| 6 | 11 30 | Clarence Yard | | |

East Goods is at Finsbury Park
Ferme Park is between Harringay and Hornsey

Clarence Yard is at Finsbury Park
M = Monday   O = only   X = excepted

### Table 5: Arrivals At and Departures From Farringdon Street Goods Depot, 14th June to 19th September 1954.

Data extracted by Steve White from the BR ER WTT King's Cross to Doncaster Main Line including King's Cross suburban district.

#### Arrivals Mon-Sat only

| No. | Time | From | Runs | Notes |
|---|---|---|---|---|
| 7 | 2 28am | East Goods | MX | |
| 2 | 5 38 | Ferme Park | MX | |
| 2 | 5 38 | East Goods | MO | |
| 6 | 10 48 | Ferme Park | | |

#### Departures Mon-Sat only

| No. | Time | To | Runs | Notes |
|---|---|---|---|---|
| 7 | 3 28am | Clarence Yard | MX | |
| 2 | 6 45 | Clarence Yard | | |
| 6 | 11 30 | Clarence Yard | SX | |
| 6 | 11 40 | Clarence Yard | SO | |

East Goods is at Finsbury Park
Ferme Park is between Harringay and Hornsey

Clarence Yard is at Finsbury Park
M = Monday   S = Saturday   O = only   X = excepted

**Note**: The first trains of the day in 1946 were numbered '1' and in 1954 were '7'. This has been checked in the original documents.

# LONDON & NORTH EASTERN RAILWAY.
## (GREAT NORTHERN SECTION.)

# WORKING TIME TABLES,

# LONDON (SUBURBAN),
## HATFIELD, HITCHIN AND BRANCHES.

# From 17th JULY to 10th SEPTEMBER, 1933, INCLUSIVE.

## FOR INFORMATION OF SERVANTS OF THE COMPANY ONLY.

### GENERAL NOTE TO PASSENGER TRAINS.

**H**  Will not convey horse boxes nor carriage trucks.

### GENERAL NOTES TO FREIGHT TRAINS.

**X**  Stop when required.     **L**  Stop for locomotive purposes only.

**†**  Shunt for other trains to pass, and not attach or detach.

When loads of Class B freight trains do not exceed Class A loads they may be run at speed laid down for Class A trains, provided the traffic is suitable to travel at the higher speed, and the letter "A" must, in such cases, be telegraphed after No. of train.

The Staff at station where alteration is made must advise Control Offices, men in charge of train, and signalmen of change. Guards must note on their journals when trains are altered as above.

The SMALL FIGURES on LEFT-HAND SIDE of train times denote the number of PASSENGER or other train for which trains are to shunt.

In cases where freight trains are not running in their scheduled times as shown in the Working Time Tables, they must run in accordance with the point to point running times laid down for the class of train concerned and Guards must account for all time lost in excess of that so shown.

WHEN ARRIVAL AND DEPARTURE TIMES ARE NOT BOTH STATED THE TIME SHOWN IS THE DEPARTURE TIME; and engine drivers and guards must ARRIVE in sufficient time to enable them to do the work AND LEAVE THE STATION at appointed times.

The time of arrival and departure of Goods Trains at Stations and Sidings are based on the normal requirements, but whenever practicable, the Trains must be got away earlier by arrangement with Control. When short of a load the Trains must run between point and point in less than booked running time.

Each Station master and Yard master must examine and compare each issue of the time books and bills (so far as his own station or yard is concerned), and report inaccuracies to Superintendent, Liverpool Street.

LIVERPOOL STREET STATION
LONDON, E.C. 2.

29th JUNE, 1933.

(2,100).

**V. M. BARRINGTON-WARD,**
Superintendent,
Southern Area (Western Section).

PRINTED BY LOVE AND MALCOMSON, LTD., REDHILL.

*Fig. 43. Cover of the LNER 1933 Working Time Table.*

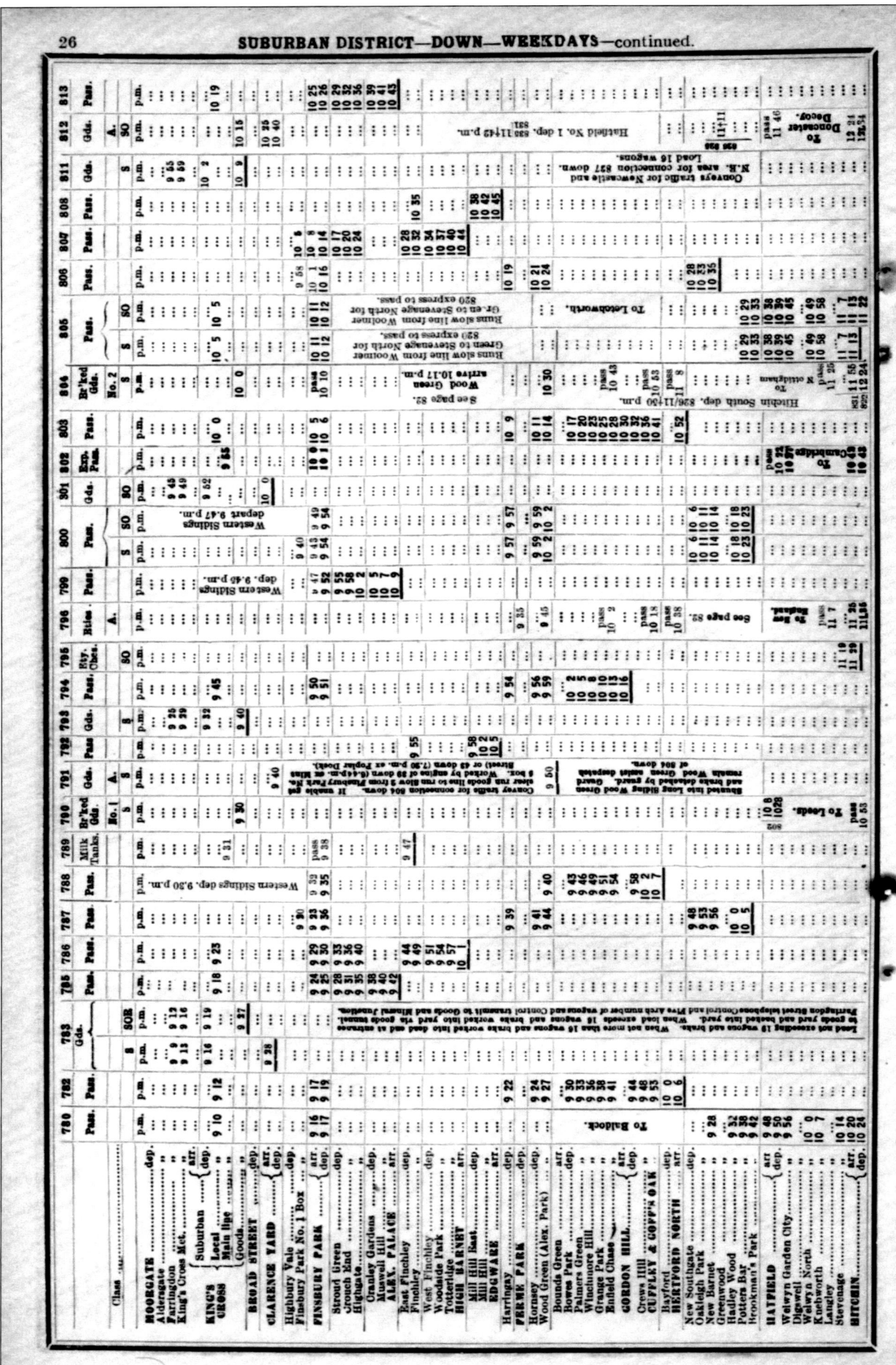

*Fig. 44.  Page 26 of the 1933 Working Time Table*

**Traffic Packed in Material such as Heather, Straw, &c.**

A number of cases have occurred in which wagons have caught fire in transit owing to sparks dropping on the packing. In future all wagons carrying loads, the packing of which is liable to ignition from this cause, must, wherever possible, be marshalled in the rear of the train. (L 31330 C.)

## TRAFFIC RESTRICTIONS ON G.N. SECTION.

**Farringdon Street Depot.**

The turntables at the G.N. Farringdon Street Depot cannot accommodate wagons with greater wheel base than 10 feet 6 inches, and wagons of a greater length than 22 feet 6 inches cannot be turned without fouling the stanchions and walls.

Wagons with a wheel base or of a length exceeding the dimensions given should not therefore be sent to Farringdon Street.

**Eight-wheel 20-ton Brakes** cannot be worked on the turntables at Farringdon Street, and they must not, therefore, be used on trains running to that station.

*Fig. 45. Extract from the L&NER (GN Section) – No. 9 Issue of Special Working Notices, in operation from 20th October 1924 UFN. It reveals that brake vans as well as goods wagons were shunted using the turntables. It is interesting to note that the arrival of 8-wheel brake vans at Farringdon Street was considered a possibility because there is very little photographic evidence of them in use anywhere on the GN system, especially the Yorkshire – London coal trains for which they were built. The same restriction would have applied at a later date to other long-wheelbase brake vans, especially LNER/BR 20-ton 'standard' vans plus those of the LMS, GWR and SR.*

## MAXIMUM LOADS EXCLUDING BRAKE VAN ON CLIMB TO LUDGATE HILL

| | Without Banker | | | | Banked to Ludgate Hill | | | |
|---|---|---|---|---|---|---|---|---|
| | Goods | Mineral | Mixed | Empty | Goods | Mineral | Mixed | Empty |
| **LNER Up** | | | | | | | | |
| To south of Thames | 22 | 15 | 22 | - | 30 | 25 | 30 | - |
| To Farringdon | 28 | - | - | 28 | - | - | - | - |
| | | | | | | | | |
| **LNER Down** | | | | | | | | |
| From south of Thames, peak hours | 25 | - | 25 | 25 | - | - | - | - |
| From south of Thames, off-peak hours | 33 | - | - | 34 | - | - | - | - |
| From Farringdon | 28 | - | - | 28 | - | - | - | - |
| | | | | | | | | |
| **LMS Up** | | | | | | | | |
| To south of Thames, 6 coupled locos. | 24 | 16 | 30 | 30 | 30 | 23 | 30 | 30 |
| 4 coupled locos. | 17 | 11 | 28 | 28 | 24 | 16 | 30 | 30 |
| To Whitecross Street, all locos. | 20 | - | - | - | - | - | - | - |
| | | | | | | | | |
| **LMS Down** | | | | | | | | |
| From south of Thames, 6 coupled locos. | 27 | 23 | 27 | 27 | - | - | - | - |
| 4 coupled locos. | 22 | 15 | 25 | 25 | - | - | - | - |
| From Whitecross Street, all locos. | 22 | - | - | - | - | - | - | - |

*Fig. 46.*

# 12 Vine Street

Although not directly associated with the GNR's Farringdon Street premises, the Metropolitan Railway's Vine Street depot was close by as can be seen on the diagram on pages 24 and 25. Small and cramped, it opened on 1st November 1909 but lasted only 17 years 8 months, closing on 1st July 1926. There are few photographs, no doubt as a result of its short life and difficult access.

*Fig. 47. The street level access on the corner of Vine Street and Farringdon Lane (London Transport Museum).*

*Fig. 48. One of the Metropolitan Railway's 1906-built BWE locomotives shunting circa 1910. These had Westinghouse control and traction equipment and bodies built by Metropolitan-Cammell (London Transport Museum).*

*Fig. 49. The same location in 2012 (Farringdon Urban Development Survey).*

*Fig. 50. Looking south-east in 2012. The large grey structure occupies the site of Vine Street goods depot. Vine Street bridge (closed) is adjacent, then Clerkenwell Road bridge. The offices on the extreme right occupy the site of Farringdon goods depot and obscure the station itself. Prominent at the junction of Clerkenwell Road and Turnmill Street is the GNR's stable block, demolished shortly after this photograph was taken and the subject of the next chapter. (Farringdon Urban Development Survey).*

# 13 Clerkenwell or Turnmill Street Stables

The Great Northern Railway employed a large number of horses for cartage and shunting in its large goods depots and took great care to ensure their well-being. The Clerkenwell stables building was constructed on the site of the Middlesex Sessions House on the corner of Clerkenwell Road and Turnmill Street and opened in 1887. Its equine career ended in the 1930s and it had several non-railway-related uses, latterly offices but with the ground floor housing a famous London nightclub called Turnmills which closed in 2008. Despite objections from English Heritage and the Victorian Society among many others the building was demolished in 2012 and was subsequently replaced by yet another modern glass and concrete office block.

*Fig. 51 (Above). An enlargement of a portion of the 1948 Aerofilms photograph on page 22 showing the Clerkenwell stables building standing apparently relatively unscathed among the damaged structures nearby at the north end of Turnmill Street. (Aerofilms)*

*Fig. 52. (Right). The Turnmill Street elevation on 4th August 2004 showing the care and attention the GNR lavished on the architectural detail and use of coloured brickwork. (Murray Hughes).*

*Fig. 53 (Above). The Clerkenwell Road [left] and Turnmill Street [ahead] elevations in 2012 with the dome of St Paul's Cathedral visible in the latter. (Turnmills).*

*Fig. 54 (Right). Even the south face of the building, half-hidden by its neighbour across Broad Yard, was afforded decorative detailing. (Murray Hughes)*

In the December 1894 issue of THE RAILWAY ENGINEER C H Grinling wrote an article about the railway companies' stables in the London area. He referred to those of the GWR, L&NWR, Midland and GNR. The March and April 1895 issues contained drawings and a description of the GNR's Clerkenwell facilities, thus:

By the courtesy of Mr Richard Johnson, M.Inst.C.E., Engineer-in-Chief of the Great Northern Railway, we publish illustrations of the magnificent new stables which were lately erected to his designs by Messrs Kirk & Randall for the Great Northern Railway Company. The building was erected on a piece of vacant ground at the corner of Turnmill St and Clerkenwell Green, a very short distance from the Gt Northern Co's Farringdon St Goods Station, for the accommodation of the horses, numbering nearly 200, employed in cartage work in connection with that depot. The premises are not, however, devoted exclusively to stable accommodation, for the basement has been converted into a warehouse for the storage of goods, while the ground floor has been laid out with platforms and carting space for the unloading and loading of vans, which can also be housed here while the horses are stabled overhead. There is also a hydraulic hoist for the conveyance of vans to and from the ground floor and the basement, when it is more convenient that these should be loaded or unloaded in the warehouse itself.

The upper part of the building, where are the stables, consists of two floors built round an open quadrangular well-hole, which is covered by a clerestory glazed roof. The stables themselves are thus built, in compartments, all round the outer walls of the building, each compartment being entered from a balcony which surrounds the well-hole on each floor. These balconies are reached by inclined approaches, having a gradient of 1 in 4. The flooring of the inclined approaches is of iron joists filled in with concrete, and covered to an average thickness of 3ins, with Wilkes Patent Metallic Paving, deeply grooved so as to hold a loose covering of peat moss. The balconies are carried on rolled iron cantilevers of the sizes figured upon the drawings, tied together at ends with channel iron. At places, however, iron brackets, securely bolted to the girders, are used in place of cantilevers to support the balconies. The entrances from the balconies to the stable compartments are provided with sliding doors hung with A Kenrick & Sons (Hadfield's patents) hanging rollers, so that all danger of the horses running foul of the doors in entering or leaving the stables is removed. At the head of the inclined approaches on each balcony water troughs are provided, constructed of galvanised cast iron in two divisions, and provided with a feed cistern covered in with sheet iron. Thus the horses can be well watered both on entering and leaving their stalls.

Each floor contains seven compartments, and each compartment from 10 to 20 stalls, and one, or sometimes two loose-boxes. The stalls are uniformly 6ft wide and 9ft 6ins in length. Including the space allowed for passages, the ground area per horse is about 90ft, while the number of cubic feet per head varies from 1,400 in the loftiest compartments to 1,000 in others. The loose-boxes average 10ft by 12ft. The stalls are all built with standing divisions,

these having been preferred to swinging bales, because of the risk, where the latter are used, of horses getting their feet caught in them, and so injuring themselves or one another. The divisions are constructed of pitch pine boarding, to which is screwed on, at the part most exposed to damage from the horses' hoofs, a kicking board of elm, 3ft square. A wrought iron capping is also provided to each stall division as a precaution against crib-biting. The loose-boxes have strong cast iron gate posts, and wooden doors also capped with wrought iron. The stall posts are of cast iron, and to each is fixed a hook for harness so that this is always ready at hand for each horse. The mangers are made of salt glazed earthenware in two pieces, each of which forms a separate trough, and can be renewed, if damaged, without disturbing the other. As only chopped food is given in these stables, no hay racks are provided. The mangers are boxed in below with boarding flush to the ground, so that there is no possibility of a horse knocking its head against the underside when rising from a recumbent position.

Each compartment is provided with two or three windows made to open and shut on the hopper principle, and furnished with Leggott's patent fanlight openers. Light and air is also freely supplied to the building from the roof covering the quadrangular well-hole, the glazed portion of which was supplied and fixed by the Pennycook Patent Glazing Co of Glasgow. The roof is raised some inches above the inside walls, and thus an aperture is provided for the constant entrance of fresh, and emission of foul, air. The whole of the walls on the balconies are faced with white glazed bricks, and the inside walls of the stable compartments are faced with straw coloured glazed bricks, to the height of the top of the stall divisions. This enables them to be washed down regularly with a hose, and contributes greatly to the cleanliness and light appearance of the interior.

The paving of the stables, balconies, and all the rooms opening off these, was laid by the Wilkes Metallic Flooring and Eureka Concrete Co, and consists of metallic paving to an average thickness of 3ins, laid *in situ*, and nicked, grooved and channelled for drainage purposes when laid down. It should be stated that a loose cover of burnt ballast was placed between this paving and the concrete in which the iron girders supporting the floors are bedded, and thus, when the girders expand or contract with changes of temperature, the loose ballast bears the strain and the pavement is preserved from cracking. For the carting space on the ground floor granite paving is provided.

For hauling up provender and bedding from the ground floor to the stables, a pulley, worked by hydraulic power, is provided in the well-hole, and from each of the two balconies is a shoot down which the manure is thrown to a chamber on the ground floor. On each floor is a large room for the storage of provender, bedding &c, and smaller rooms for storing and repairing harness. On the ground floor is a dwelling house for the chief stable keeper, a shoeing forge in which three men are constantly at work, and mess room for van men. Another mess room for the horse keepers is provided on the top storey.

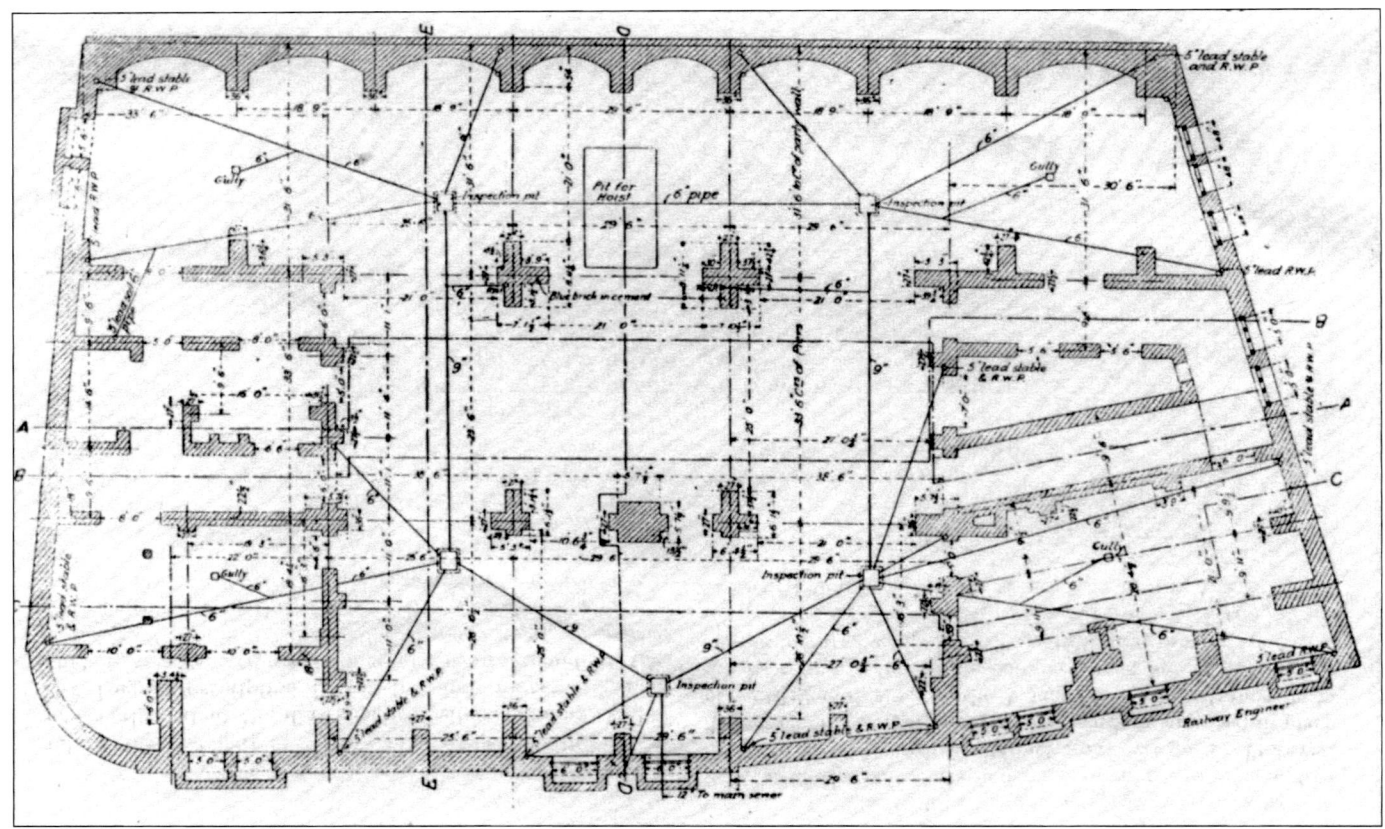

*Fig. 55. Plan - basement, drains, &c*

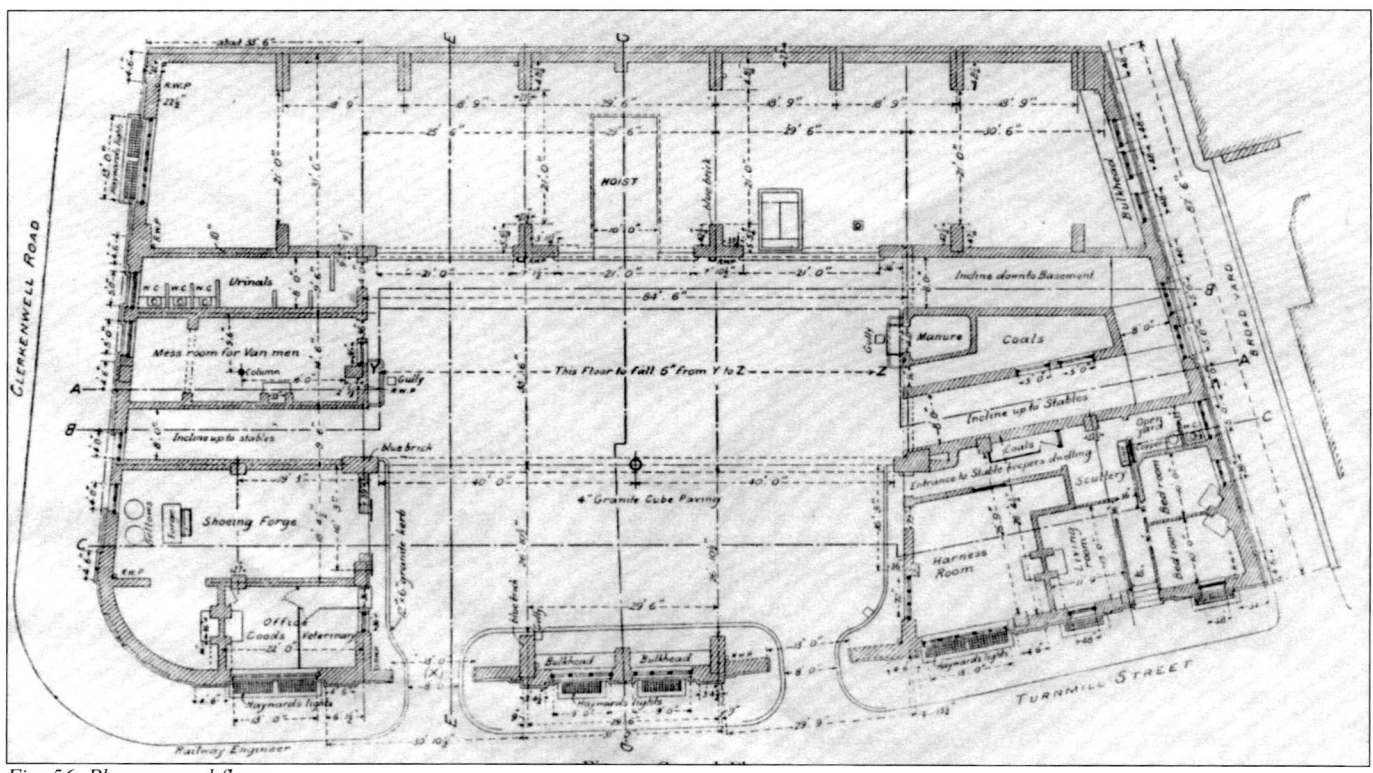

*Fig. 56. Plan, ground floor*

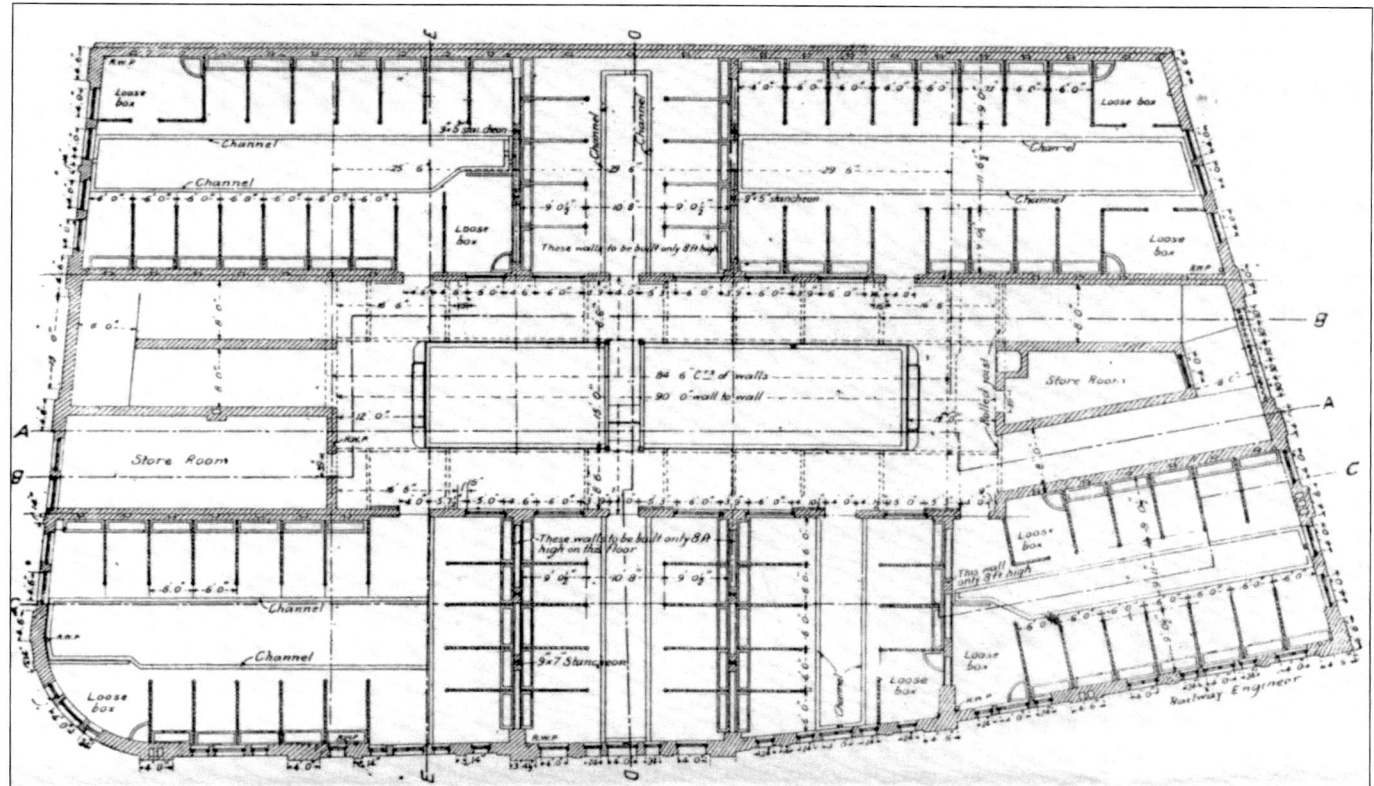

*Fig.. 57. Plan, first floor*

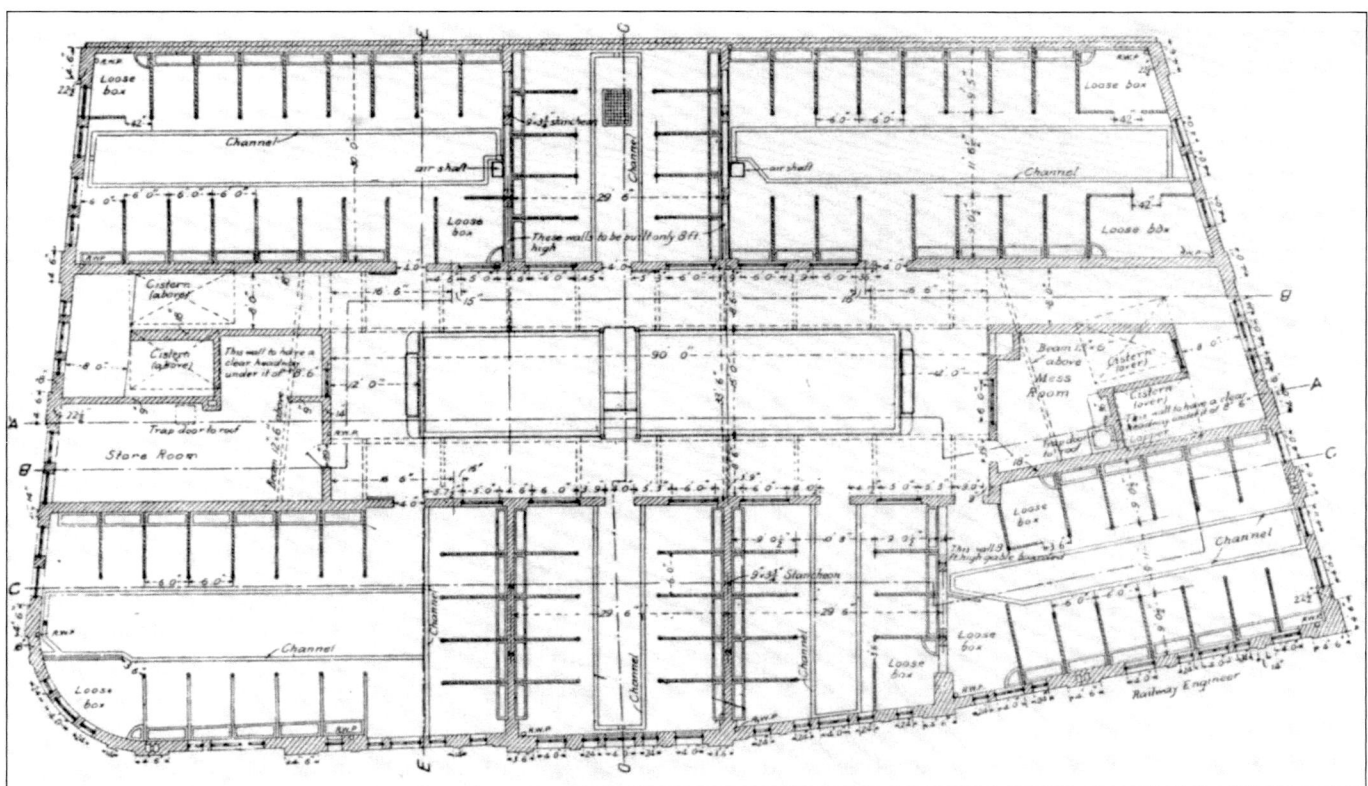

*Fig. 58. Plan, second floor*

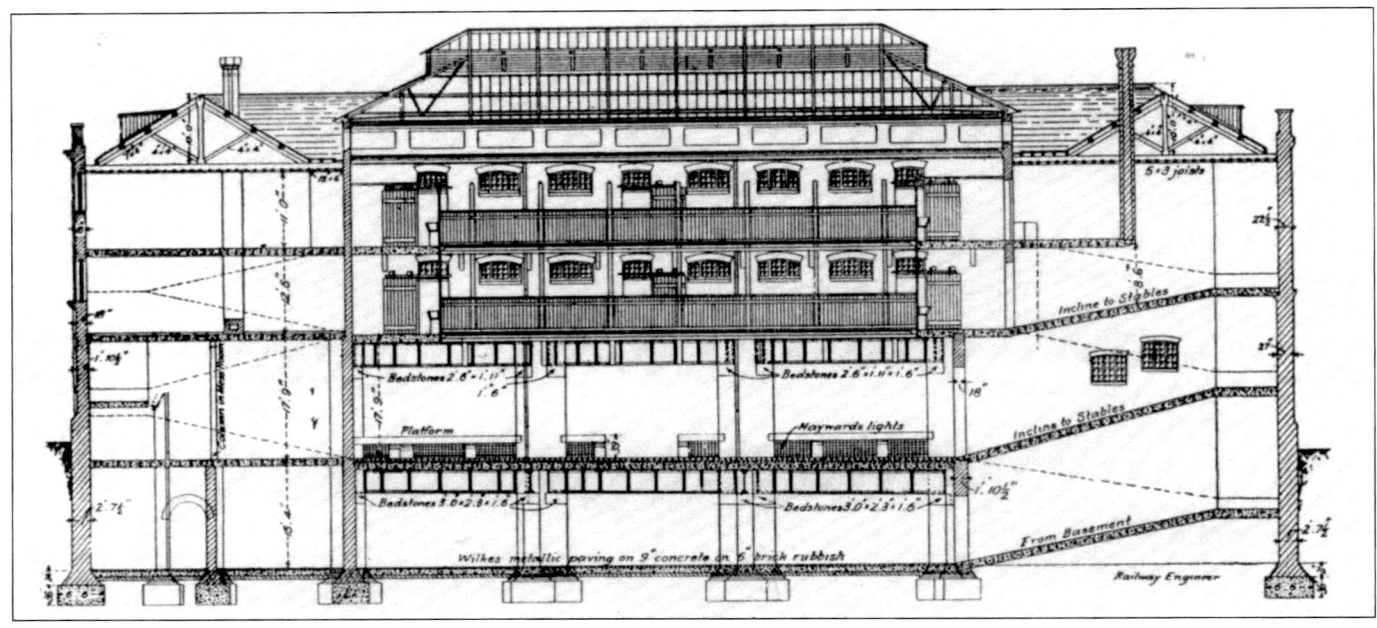

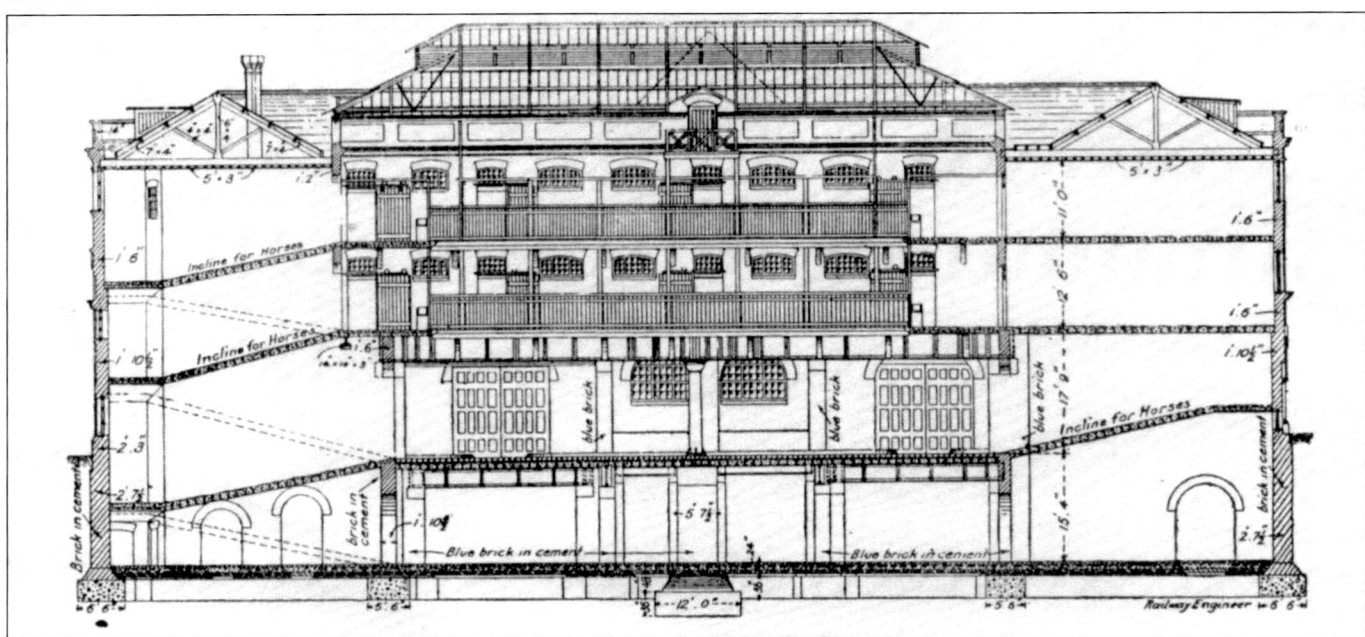

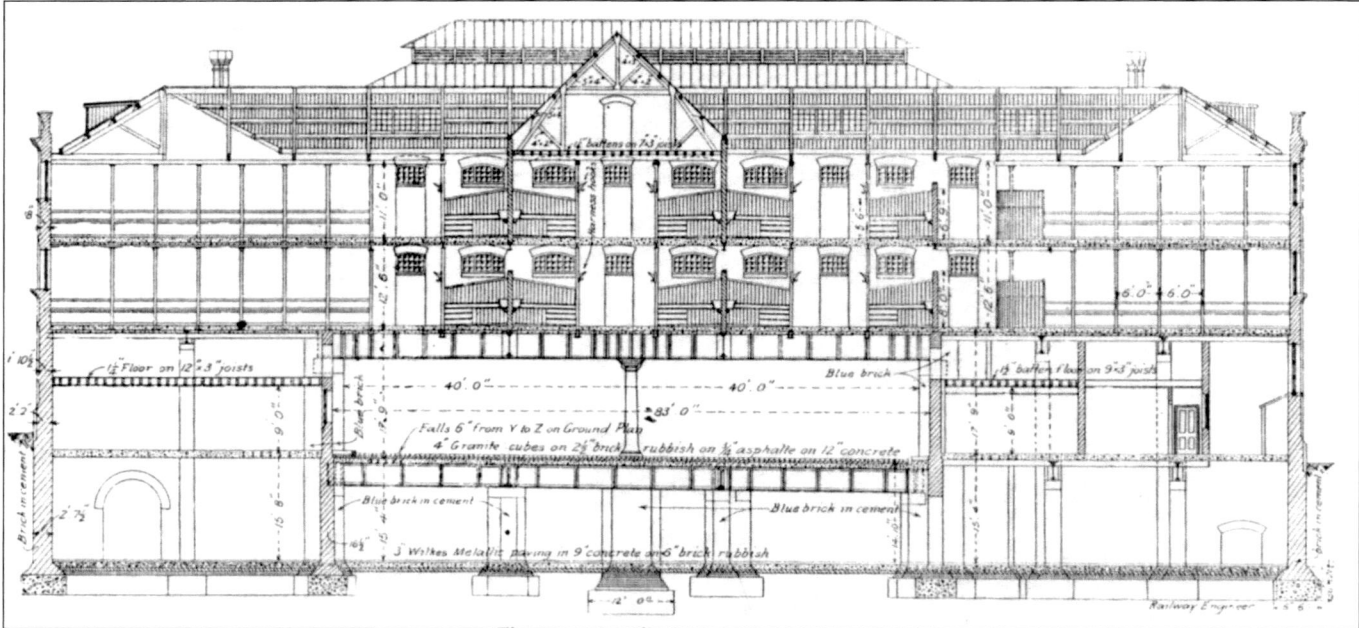

*Figs. 59, 60 and 61. Longitudinal sections on plans (figs. 56-58) : A-A (top), B-B (centre) and C-C (bottom)*

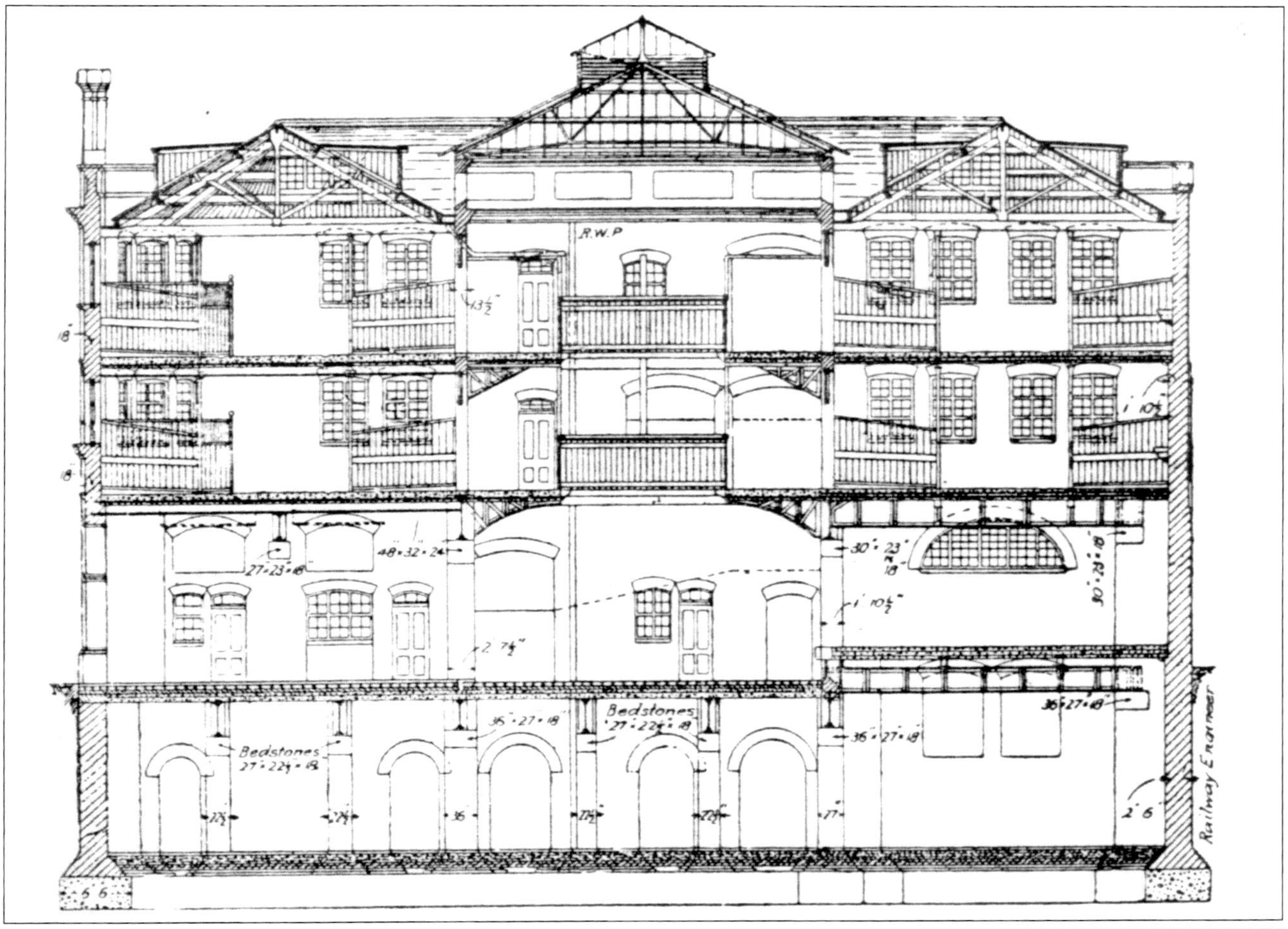

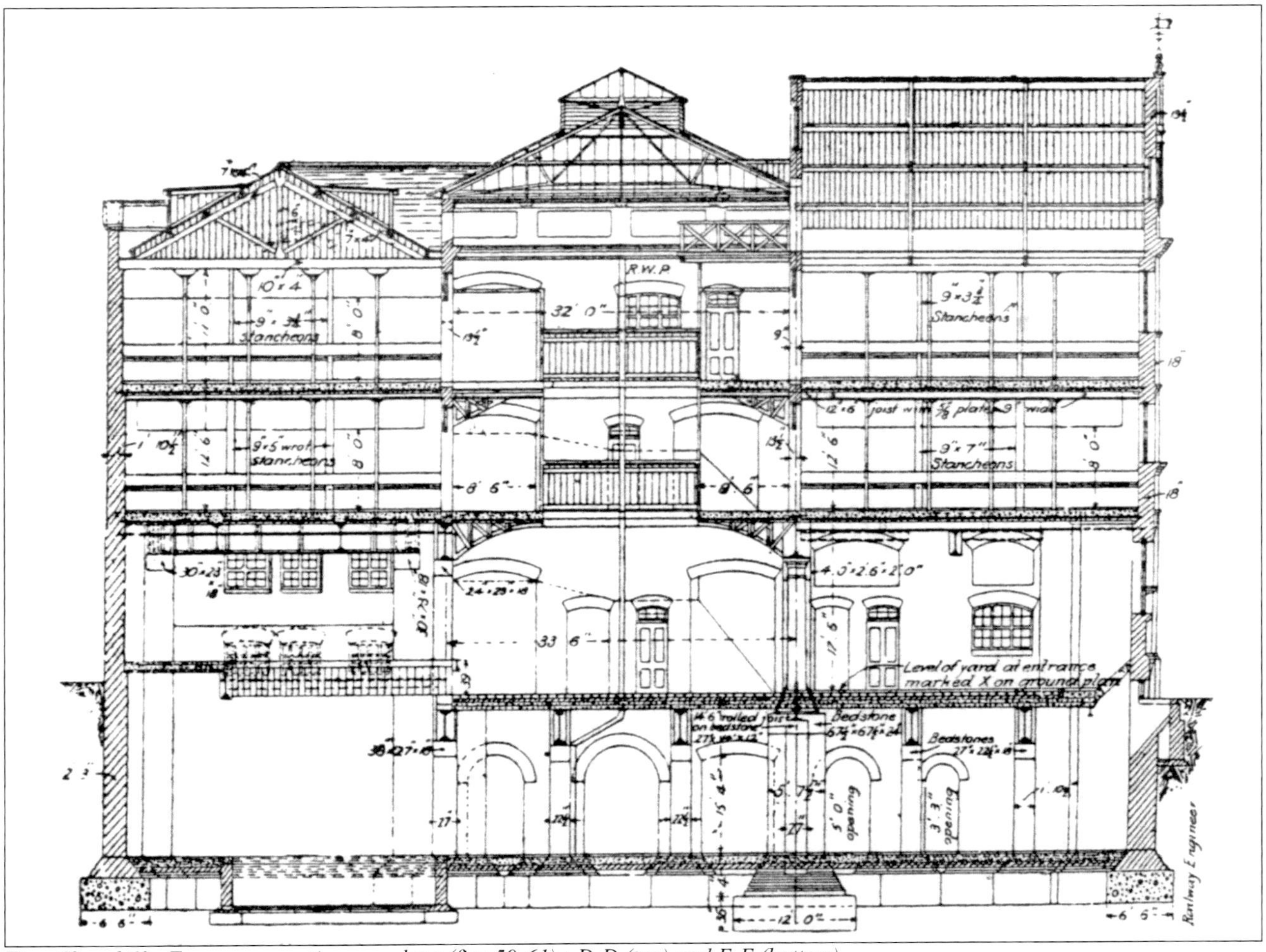

*Figs. 62 and 63. Transverse sections on plans (figs 59-61) : D-D (top) and E-E (bottom)*

# 14 Taking Care of Railway Horses

The following extract, covering only those sections relevant to the GNR, has been transcribed from pages 248 - 258 of a book THE WAYS OF OUR RAILWAYS by C H Grinling, published in 1910. The accompanying photographs are not of the greatest quality having been scanned from a book printed some 110 years ago but they are included for illustration.

On the GNR horses outnumber loco-motives by about two to one, the company having 2,782. The Midland has over 5,000, and the L&NWR nearly 6,000. Most railway companies pay a good price for their horses, buying them young and in the best of condition after they have been well broken in to work by a spell on a farm. The average working life of a railway horse is between five and six years, at the expiration of which time the animals still have sufficient work left in them to command a fair price for farm duty again. About £60 is considered a fair price to pay for a recruit for a railway stud, and one of the companies gets on average as much as £23 for each horse sold after nearly six years of regular service. This last figure is, however, rather above the average price realised on horses sold by the companies generally. Usually out of a large railway stud three or four at least are killed by accident during a year, while the number that die or have to be destroyed is about two per cent of the total kept. Even these, however, are not a total loss, as the carcasses sell for about thirty shillings each.

The Great Northern at Clerkenwell has three storeys of stables accommodating 189 horses in a building the ground floor of which is devoted to a goods warehouse. The animals make no difficulty of ascending and descending the inclined ways which lead from one floor to another, the fodder and bedding being taken up by mechanical hoists. For the erection of its newest stables the Great Northern has utilised a piece of waste land situated over the entrance to one of its London tunnels, and here nearly 200 horses are accommodated in a handsome triangular range of one storey buildings. This company has stables underground, on the level, and with several floors, and its horses thrive equally well in each of these types of home.

There is then a section devoted to detailed descriptions of bedding, provender and its production, shoeing, harness making, etc. They are not specific to the GNR and the item on provender is based on Grinling's observations during a visit to the GER provender manufactory and store at Romford, although he mentions GNR provender store at Holloway. He then goes on to describe the GNR horse hospital, thus....

HOSPITAL STABLES, NEAR KING'S CROSS, G. N. R.
Photo by C. Pilkington.
*Fig. 64*

Notwithstanding the care taken in respect to food and stabling, and the fact that all the animals are in the prime of life, sickness amongst railway horses cannot be entirely prevented.

The severity of their work, relieved though it be by frequent periods of rest, makes them specially liable to ailments of the feet and legs, whilst throat and lung troubles are also common, especially with those which work in towns. It is a noteworthy fact that almost all the young horses develop colds as soon as they arrive in town from the farms at which they have been bred.

To provide accommodation for "seasoning" these recruits and for the treatment of the sick, every large railway company has its special range of hospital stables at its headquarters, and most of them also have a country convalescent home within easy distance of their principal horse depot. The Great Northern's horse hospital - to again give one example typical of the rest - is situated within a mile of King's Cross, on an elevated breezy site at the summit of that remarkable region north of the Euston Road, which is given up almost entirely to the occupation of the three great trunk lines running from London to the north. Miles and miles of sidings - the existence of which passengers by these lines know next to nothing - testify to the immensity of the traffic in goods and coal which the needs of the Metropolis demand. One can imagine that the young farm horse, when discharged from the "box" in which he has performed his journey to town, feels as great a shock of surprise at the change in surroundings as does the country bumpkin on emerging into the crowded streets from the great passenger termini near by.

A few weeks' rest in one of the quiet, cosy stalls of the "reception stables" under the daily care of an experienced vet soon sets him on his "town legs". Indeed, this horse hospital in the heart of busy London is in some respects the counterpart of the farm from which he has come - so fragrant are the country smells, so homelike the look its straw-yard and its ranges of well-filled stalls. Equally refreshing must this rural atmosphere be to the invalids - gathered here from all parts of the Great Northern system - who occupy a homely looking two-storey range of stables forming the other wing of the hospital. Here every resource of veterinary science is at the service of the sick animals. Even oxygen and chloroform are not denied them, while the arrangements for hydropathic treatment, if not so elaborate as at Matlock or Ilkley, are an efficient aid in the treatment of foot troubles.

Even the "Turkish bath" is sometimes called into requisition for sweating sick horses and at Totteridge, where the Great Northern has its convalescent home for horses, this was, until recently, part of the regular treatment in certain cases. Modern veterinary opinion, however, favours hot baths of the "Russian" variety, it having been found that horses sweat more freely under the application of steam than they do in a hot air chamber.

A HORSE AMBULANCE, G. N. R.

*Photo by C. Pilkington.*

*Fig.65*

provender sacks were being turned out in a day, as compared with fourteen a day before its introduction.

In another room a man was engaged in making horse collars. This is a highly skilled form of work that requires a long apprenticeship, but a skilled operator can turn out seven in a week. As showing the care exercised to prevent waste, it is interesting to know that those parts of the stuffing used for railway carriages which would otherwise be thrown away are sent to the horse department for use in the making of collars. To prevent the horses' necks becoming galled, the collars in use are dried after each day's work. Each horse has its own collar set apart for it, so as to ensure an easy fit. No detail is thought too trivial, if it conduces to the health of the animals or to economy of work in the department.

Shoeing forges and harness-making shops are interesting auxiliaries to the horse departments of our railways. At small stations the shoeing is let out to contract, but at most large depots the railway companies find it more economical to have their own forges. As to harnesses, horse collars, and nosebags, the practice varies, but some companies make all these articles in shops of their own, with the aid of the most approved types of labour-saving machinery.

At King's Cross I saw a stitching machine in the harness shop, driven by electric power supplied from the lighting works at Holloway, which did in twenty minutes the amount of work which it would have taken a man a day to do by hand. Near by was another machine, with the help of which eighty

Grinling then goes on to comment on the traffic chaos in central London caused by innumerable goods vehicles parking in the streets in front of commercial premises for loading and unloading. Yes, that was in 1910! According to Murray Hughes the "newest stables" to which Grinling refers were by the south portal of Copenhagen tunnel, and later rented out by the LNER to Cadburys of York.

HORSES BATHING AT G. N. R. HOSPITAL, NEAR KING'S CROSS, LONDON.

*Photo by C. Pilkington.*

*Fig. 66.*

# 15  References and Acknowledgements

**References:**

-. A TO Z ATLAS OF LONDON, Geographers' Map Co Ltd (various undated editions)

-. THE ENGINEER, 23rd November 1894 pp 440-444

-, THE RAILWAY ENGINEER, December 1894, February 1895 and March 1895 (articles attributed to Charles H Grinling)

Borley, H V.  CHRONOLOGY OF LONDON RAILWAYS, R&CHS 1982 ISBN 0901461334

Brooksbank, B W L.  LONDON MAIN LINE WAR DAMAGE, Capital Transport Publishing 2007, ISBN 9781854143099

Day, John R.  THE STORY OF LONDON'S UNDERGROUND, London Transport 1963, ISBN 0853290946

Goslin, Geoff.  STEAM ON THE WIDENED LINES, VOLUME 1, Connor & Butler 1997, ISBN 0947699252, Volume 2, ditto 1998, ISBN 0947699287

Grinling, Charles H. THE HISTORY OF THE GREAT NORTHERN RAILWAY, (1) Methuen, 1898, 1903 and
(2) George Allen & Unwin, 1966

Grinling, Charles H. THE WAYS OF OUR RAILWAYS, Ward, Lock & Co, 1905

Hilliam, David. WHY DO SHEPHERDS NEED A BUSH? THE HISTORY OF LONDON UNDERGROUND STATION NAMES, The History Press 2010, ISBN 9780752455266

Hunter, Michael and Thorne, Robert.  CHANGE AT KING'S CROSS, Historical Publications Ltd 1990, ISBN 0948667060

Huntley, Ian.  LONDON UNDERGROUND SURFACE STOCK PLANBOOK 1863-1959, Ian Allan 1988, ISBN 0711017212

Jackson, Alan A.  LONDON'S METROPOLITAN RAILWAY, David & Charles 1986, ISBN 9780715388396

Sibley, Allan (Editor). GREAT NORTHERN NEWS No. 117 pp 14-22 "Royal Mint Street" (various contributors), Great Northern Railway Society 2001

Smith, Martin.  STEAM ON THE UNDERGROUND, Ian Allan 1994, ISBN 0711022828

Temple, Philip.  'FARRINGDON ROAD', SURVEY OF LONDON VOL. 46 (SOUTH AND EAST CLERKENWELL), English Heritage 2008
published at http://www.british-history.ac.uk/report.aspx?compid=119247.  Includes reports in "The Builder", 9 May 1874, p.397; 16 May 1874, p.425; 19 June 1875, pp.549–50; 26 June1875, p.586; 4 Sept 1875, p.806; 21 Sept 1878, pp.995–6 plus TNA, RAIL 236/522

Wrottesley, John.  THE GREAT NORTHERN RAILWAY (three volumes), Batsford,
Vol I, 1979 ISBN 0713415908,  Vol II, 1979 ISBN 0713415924, Vol III, 1981 ISBN 0713421835

Young, John N.  GREAT NORTHERN SUBURBAN, David & Charles 1977, ISBN 071537477X

**Acknowledgements**

I am grateful for the help given by the following:

Staff at the National Railway Museum, York.
Staff at the London Transport Museum, Covent Garden, London.

| | | | | |
|---|---|---|---|---|
| Dave Backhouse | Jonathan David | Adrian Marks | Mike Perrins | Paul Taylor |
| David Cockle | Murray Hughes | Adam Muir | Marion Sibley | Steve White |
| Jim Connor | | | | |